Inventing the Modern Self and John Dewey

Inventing the Modern Self and John Dewey

Modernities and the Traveling of Pragmatism in Education

Edited by Thomas S. Popkewitz

INVENTING THE MODERN SELF AND JOHN DEWEY
© Thomas S. Popkewitz, 2005.

First published in 2005 by
PALGRAVE MACMILLAN™
175 Fifth Avenue, New York, N.Y. 10010 and
Houndmills, Basingstoke, Hampshire, England RG21 6XS
Companies and representatives throughout the world.

PALGRAVE MACMILLAN is the global academic imprint of the Palgrave Macmillan division of St. Martin's Press, LLC and of Palgrave Macmillan Ltd. Macmillan® is a registered trademark in the United States, United Kingdom and other countries. Palgrave is a registered trademark in the European Union and other countries.

ISBN 1–4039–6862–4

Library of Congress Cataloging-in-Publication Data

Inventing the modern self and John Dewey : modernities and the traveling of pragmatism in education / edited by Thomas S. Popkewitz.
   p. cm.
Includes bibliographical references and index.
ISBN 1–4039–6862–4
   1. Dewey, John, 1859–1952—Influence. 2. Education—Philosophy.
I. Popkewitz, Thomas S.

LB875.D5159 2005
370′.1—dc22   2005048682

A catalogue record for this book is available from the British Library.

Design by Newgen Imaging Systems (P) Ltd., Chennai, India.

First edition: December 2005

10 9 8 7 6 5 4 3 2 1

Printed in the United States of America.

Contents

Preface

John Dewey remains today as a heroic figure whose internationally circulated ideas about pragmatism offered a way to think about a progressive individual associated with modernity. Born in 1859 and dying almost a century later, Dewey wrote across the domains of political theory, philosophy, psychology, and education, but also as a public intellectual. He was a quintessential American thinker of his time, a time in which the nation he "represented" rose to the peak of international significance. Dewey lectured about pragmatism and his books appeared in such diverse places as Australia, Brazil, Britain, China, Germany, Japan, Finland, Serbia, Turkey, and Sweden. Yet at the same time, Dewey's pragmatism became an archetype of a form of modernism that religious, intellectual, and political elites rallied against in places such as Brazil, China, Germany, and the former Soviet Union. Today, Dewey's writings are viewed as instructive for thinking about contemporary social life and education's challenges.

The purpose of the volume is not to present Dewey as an icon of intellectual or educational thought nor as the messenger of pragmatism. While I do not want to detract from the particular contributions of Dewey to philosophy and education—his position as an author— this volume historizes that authorship to consider the issues of governing of the "self" presupposed in the productions of modernity and the "modern" school. Dewey's pragmatism was one of many cultural theses that traveled internationally in a field of multiple ideas, authority relations, and institutions about the moral and spiritual values governing the constructions of the child. He "participates" in salvation narratives of the Mexican Revolution, Pan Slavic identity in Yugoslavia, to Chinese reform or challenges to Confucius hierarchies school reforms. Further, Dewey signified the enemy of spiritual values of the nation as counter cultural theses of modernization in the reform of schools were sought in Brazil during the early twentieth century and by Mexican leftists and Maoists in China later in the century.

The essays, then, do not focus on the international and national debates about the new education and schooling but the intersection of pedagogical projects with broader cultural, national, and institutional patterns about the constructions of modernity and the modern "self" during the twentieth century. The historical approach places John Dewey and pragmatism in a paradoxical position. Dewey enters an international field concerned with issues of modernization through the school, but this position is provincialized and decentered in this book to explore different cultural projects of the school in reforming who the child is and should be. Dewey's pragmatism traveled internationally, provoking both admiration and fear as global changes brought a re-visioning of the cultural narratives about who the citizen and child are. Dewey traveling—figuratively and literally—enters different cultural and social flows and cultural narratives about society and individuality in the making of the New Education, a term that circulated in the first decades of the century. This traveling continues in current international discussions about constructivist approaches to pedagogy.

In addition, studies of Dewey and pragmatism have focused on either questions of intellectual history or social history, in particular as it relates to pragmatism as a "uniquely" American philosophy or how others borrow the ideas of Dewey in a particular country. This volume, in contrast, studies Dewey to shed light on the heterogeneous processes through which modernity and modern individuality are constructed. This book brings together scholars from whose studies of nine nations provide original chapters that place Dewey within an international field in which "modern" schools in its plural sense emerge during the twentieth century.[1] Its scholarly contribution crosses multiple fields: studies of Dewey and pragmatism, educational studies of the modern school, histories related to modernity, studies of globalization, and cultural studies concerned with knowledge and the problem of governing.

Two key ideas organize the studies in this book about Dewey and the construction of modernities.

First is the ironic phrase *the indigenous foreigner*. The phrase renders the global and the local intelligible by exploring the global circulation of particular epistemic qualities of Dewey. The ideas and concepts of Dewey, for example, are reassembled in pedagogical discourses of the Balkans, Brazil, Mexico, and Japan to produce new spaces of collective belonging and "home."

Second, the "thought" of Dewey is part of a *traveling library*, a term I borrow from Marc Depeape. Dewey's ideas circulated with other American pragmatists such as James, and were invited into conversations with French and German pedagogues in the Balkans, Belgium, Turkey, and Mexico to express national and social aspirations about the new education, the conduct of the individual, and the "nature" of the citizen. At other historical moments and places, the notions of pragmatism exemplified in Dewey were categorized as the enemy of progress and harmony in society and the nation such as in Germany or Brazil.

This analytic approach to Dewey as part of a traveling library considers how ideas are constructed in a field of "global" and local authors whose resultant patterns of "thought" are not merely the sum of its parts or a variation of a constant theme. The assembled libraries inscribe patterns given to the modernizing of the nation and the individual via some of the words and concepts of Dewey, but these patterns are not necessarily traceable to Dewey as the original author. Dewey in this volume, then, is not one "person" but the connections and disconnections produced as principles to order different modernities and one of its major institutions in the fashioning of the citizen, the school.

The historical approach to Dewey as an indigenous foreigner is to counter what Walter Benjamin has called an *emptying of history*.[2] That empty history has Dewey appearing as a logical system of thought or "concepts" that has no social mooring in the interpretations and possibilities of action. This is most evident in studies of Dewey where he is treated as an icon of reform and progressive thought without exploring the epistemic and cultural history in which that "thought" is made possible and intelligible. The cultural and social history of this volume explores the multiple flows and amalgamation of practices (traveling libraries) through which "thought" is produced. Approaching the study of Dewey in this manner is to recognize that pragmatism is one of the multiple systems of "thoughts" and cultural thesis in what is called modernity(ies).

The book brings into focus a number of important themes in contemporary philosophy, social science, history, and education. Among these are:

(1) The collection of essays provides a way of rethinking the history of schooling from a comparative and cultural/social historical perspective. It allows one to consider the field of ideas, authority relations, and institutions through which modern schools are constructed, and the multiple principles of its pedagogical practices in governing who

the child is and should be. The essays are not normative or celebratory but analytical and diagnostic of the cultural, social, and political processes through which the child is constituted as an object of reflection and action. The tracing of the movements of pragmatism, then, is a way to consider the principles in governing the principles of reflection and participation—schooling and its constructions of the child.

(2) It is within this intellectual project that the examination of Dewey's pragmatism is placed. Pragmatism functions as a cultural thesis about modes of living. That mode of living is one of many cultural theses about how individuals should reflect and participate—the modes of living—that circulate in the twentieth century. The significance of exploring these cultural theses is they also embody norms and values of a collective belonging and "homes" through which individuality, society, and the nation are constructed.

(3) The essays of this book place the construction of modernity within a field of cultural practices in which European and North American contexts are viewed relationally and historically. The assumption of the essays is that modernity is not a fact to explain the changes occurring but a term whose use is made possible through changes in ideas, authority relations, and institutions. In addition, these changes are best understood internationally and relationally, thus my use of the "modernities" in the title. The emphasis of multiple modernities as embodying different cultural theses about how one should live and act is to recognize that the knowledge that we have of our "selves" is not merely an epiphenomenon of other institutional or structural changes, but an integral and material element of governing.

(4) The book provides a way to study new expertise that relates political and elite knowledge with a popular knowledge of everyday life. The different discussions, for example, focus on the inscription of science as a way of organizing daily life rather than as a way of studying the natural world. This inscription of science as a set of principles for living, however, was not only about rational decision-making and paths for solving problems. The rationality to order "thought" embodied new values and norms that constitute the sublime through which one may think and act in daily life. Thus, there are different notions of beauty and the good life that overlap with notions of science in Switzerland and Columbia, to use two examples of the book, as well as the inscription of fear about the characteristics of the child that hinders progress. Further, the essays provide a way to understand how knowledge itself— its ways of constituting the objects of reflection and participation—are governing practices formed through the theories and methods of new social and educational sciences.

(5) The historical approach of the book provides a strategy for studying the relationship between the global and local in the practices of constructing modernities. There is an industry of work on globalization that tends to focus on the economic, political, and social, while ignoring that globalization is not something only of the present but something that needs to be considered in how the present is different from the past. This historical differentiation is part of the discussions that take place in this book. In addition, the focusing on different modernities through its cultural theses can provide historical lenses of the present, particularly in thinking about current social and political conflicts that entail debates related to religion, political regimes, and cultural regulations.

The book overlaps with multiple intellectual themes and disciplinary foci and thus has multiple reading publics. The chapters conceptualize the history of schooling and its politics of knowledge in an international field from which modern schooling emerged. In this sense, the collection of chapters provides both a substantive and methodological intervention in the study of schooling. The chapters also respond to renewed interest in John Dewey and pragmatism that moves across the terrains of education, philosophy, and social theory. At the same time, the book brings together and integrates historically disparate discussions that relate to globalization and localization, multiple modernities, and issues of governing and government. The latter is not about the formal properties of the legal administrative apparatus of the state but the governing that occurs through schooling in generating principles about who the child and citizen are and should be.

THOMAS S. POPKEWITZ
Madison, WI

Notes

1. The introductory chapter discussion places Dewey as a conceptual personae in the United States as related to the contexts of the other chapters.
2. W. Benjamin, ed., *Illuminations: Essays and Reflections* (New York: Schocken Books, 1955/1985).

Acknowledgments

The writing of this book involved conversations with many different people whose kindness made its development possible. The book was initially discussed with Lynda Stone as one of an interaction of philosophers and historians and as an outgrowth of a symposium that Lynda organized at the American Educational Research Association meeting using my notion of indigenous foreigner to discuss John Dewey. As the book progressed, it took on a more historical approach to explore the field in which Dewey, pragmatism, and modernism overlapped. But the conversations with Lynda were important in pushing my thinking about Dewey as an *historical personae*, which I talk about in chapter 1. I also have to acknowledge the contribution of the writers of this book as I tried to think about the historical specificity of Dewey in an international setting. I am not sure that they bargained for this when writing their chapters, but I appreciated their continual willingness to let me bombard them with emails full of questions.

In writing chapter 1, my debt goes to these contributors as well as to a number of people who were gracious in looking at what I wrote and in helping me think further about the problem and problematic of the book. I thus must thank the following: Matt Curtis, Cleo Cherryholmes, Inés Dussel, Ruth Gustafson, Jamie Kowalczyk, António Nóvoa, Miquel Pereyra, Amy Sloane, Noah Sobe, Mirian Warde, and Dar Weyenberg. In different ways, each provided responses to my questions that continually pushed my thinking, although not always agreeing with my conclusions. Ghita Steiner-Khamsi and Ana Isabel Madeira offered their historical understanding of African educational reform for thinking about Dewey, which I discuss briefly in chapter 1.

There are also places and people that helped to get this project started and working. "The Research Community Philosophy And History Of The Discipline Of Education: Evaluation And Evolution Of The Criteria For Educational Research," established by Marc Depaepe and Paul Smeyers with a grant from The National Fund for

Scientific Research—Flanders, Belgium (Fonds voor Wetenschappelijk Onderzoek—Vlaanderen) is a small yearly international seminar where an ongoing space for discussion is provided. Started in 1999, my discussions with the historians and philosophers at these meetings are central in the formulation of this book, with many of its participants becoming contributors. The 2003 sponsorship of a conference in Leuven, Belgium enabled many of the authors to discuss the focus of this book.

Finally, the completion of this project was made possible by the Finnish Academy of Science fellowship to The Helsinki Collegium in the fall of 2004. I appreciate the invitation of Tuula Gordon, the Associate Director and a cultural sociologist who arranged for my fellowship, and its Director Juha Sihlova who ensured the necessary academic support. Taina Seiro and Maria Soukkio deserve special thanks. They had the responsibility of making sure that the fellows quickly got settled into the office and the university-owned apartment, and could start to work with little or no disturbances. They were successful. The architectural pleasures of the center of the City of Helsinki and its people, and Stockmann's *Hullut Päivä* made the rest easy. Finally, I want to thank Amanda Johnson, my editor at Palgrave and Eva Talmadge who followed through on the production. Finally, there are Christine Alfery who helped to design the cover with the photo from Alhambra for this book as well as to push me to think about what it provokes in its imagery, and Chris Kruger of my department whose secretarial and general organizational skills keep my scattered existence and this book together.

List of Contributors

Sabiha Bilgi is a doctoral candidate at University of Wisconsin-Madison, United States. Her research interest lies in historical and cultural studies of childhood and education, with special focus on Turkey.

Rosa N. Buenfil Burgos is Professor at the Departamento de Investigaciones Educativas, Centro de Investigación y Estudios Avanzados, México. Her research brings together historical and political philosophical approaches in studies of educational discourses and issues of globalization in Mexican educational policies.

Marc Depaepe is President of the newly founded Faculty of Psychology and Educational Sciences at the Katholieke Universiteit Leuven, Campus Kortrijk (KULAK) and Professor of History of Education at the Katholieke Universiteit Leuven, Belgium.

Tom De Coster is a scientific collaborator at a joint research program between the Ghent University and Katholieke Universiteit Leuven, Belgium.

Jorge Ramos do Ó is Professor of History of Education at the Faculty of Psychology and Education Sciences at the University of Lisbon, Portugal. He has written about political history, history of culture and mentalities, during the authoritarian period in Portugal (1926–1974), and also about the history of education and modern pedagogy (nineteenth and twentieth centuries).

Kentaro Ohkura is an Assistant Professor of Comparative Studies and Social Theory of Education at Tamagawa University, Tokyo, Japan. His current interest lies in reexamining curriculum studies from historical and cultural contexts.

Ulf Olsson is Assistant Professor of Science of Education at Stockholm Institute of Education, Sweden. He is currently working in a project: *The State, the Subject, and Pedagogical Technology: A Genealogy of* *the Present of Political Epistemologies and Governmentalities at the Beginning of the Twenty-First Century.* His main research interest concerns the fabrication of subjects and governmentalities in the field of Public Health.

Seçkin Özsoy teaches at Ankara University, Department of Educational Sciences, Turkey. He completed his doctoral dissertation in 2002 on the problematic of equality and rights in higher education. His research focuses on educational rights discourse and the relation of education–ideology–hegemony.

Kenneth Petersson is Associate Professor in Communication and senior lecturer at Linköping University, Department of Thematic Studies, Sweden. His research interests include the political government of the educational field and criminal justice in Sweden. Currently he is studying Swedish mentalities that govern contemporary strategies of pedagogical thought and technologies in different discursive and institutional practices.

Thomas S. Popkewitz is a Professor in the Department of Curriculum and Instruction, The University of Wisconsin-Madison, United States. His research concerns the politics of knowledge and the systems of reason that govern policy, pedagogy, and research.

Jie Qi is an Associate Professor of Education and Japanese at the Utsunomiya University, Japan. Her research interests include postmodernism theories in teacher education and multicultural education, educational philosophy, and political sociology of education.

Javier Sáenz-Obregón is Associate Researcher of the Centro de Estudios Sociales of the Universidad Nacional de Colombia in Bogotá. He has published extensively on the history of pedagogical practices, the nonrational dimension of pedagogy and educational and cultural policy.

Frank Simon is Professor in History of Education at Ghent University, Belgium. For the last decade his research has focused on everyday educational practice, classroom and curriculum history. He is editor-in-chief of *Paedagogica Historica. International Journal of the History of Education* and is a member of the Executive Committee of ISCHE.

Noah W. Sobe is Assistant Professor at Loyola University Chicago in the Cultural and Educational Policy Studies program. He researches the history of education in East/Central Europe and is interested in student and teacher travel and questions of mobility as they intersect with the historical formation of curriculum and pedagogy.

Daniel Tröhler is Professor of Education and Head of the Pestalozzianum Research Institute for the History of Education at the Zurich University of Applied Sciences: School of Education, Switzerland. His research interests include the history and historiography of education, semantical analysis of politico-educational languages, republicanism and education, and the internationality of research.

Angelo Van Gorp is Assistant Professor of History of Education at the Katholieke Universiteit Leuven, Belgium. He has special interests in the history of educational sciences, progressive education, special needs education, and child care, which flows from his doctoral dissertation on the Belgian psycho-pedagogue Ovide Decroly (1871–1932).

Mirian Jorge Warde is Professor at the Pontifical Catholic University of São Paulo, Brazil. Her research focuses on history of education and culture, especially on the processes of internationalization–nationalization of the educational and pedagogical patterns in Brazil and United States. She has written intensively about the late nineteenth century and the first half of twentieth century.

I

Introduction

I

Inventing the Modern Self and John Dewey: Modernities and the Traveling of Pragmatism in Education—An Introduction

Thomas S. Popkewitz

Introduction

This book explores the relationship of education to modernity, pragmatism, and one of pragmatism's major international spokespersons, John Dewey. As I look across a number of social science and philosophy journals today, these three themes have found a certain prominence in contemporary scholarship for thinking about the problems, issues, and dilemmas of the present.

The idea of modernity, for example, is used in different ways to refer to changes of the past and the present. Various debates in the social sciences, history, and philosophy about what to label the present—modernism, postmodernism, advanced modernity, to name just a few—suggest a certain unease with the present and its constructions of historical legacies and present coherences. For the most part, discussions about modernity signify it as a noun, a particular object explained by a range of changes in the *long* nineteenth century[1]— social and economic, scientific inventions, and new technologies (railroad, electricity, and telegraph), industrialization and urbanization, as well as the new institutions of the state concerned with the welfare of its citizens. Others go back to the eighteenth century to talk about the emergence of a particular intellectual outlook that forms with, for example, the Reformation and Counter-Reformation, Enlightenment thinkers, and new political regimes such as those associated with the American and French Revolutions.

This chapter focuses on the educational theories about teaching and the child (pedagogy) as they relate to the constructions of modernity and the "modern" child during the twentieth century. The starting point of the discussion is with John Dewey and pragmatism. The distinctly American approach for reforming society through reforming the child in the school entered into an international field where related issues of changes were being underwritten about society and schooling. Dewey's pragmatism traveled to China, Japan, Turkey, Brazil, Belgium, and Sweden, examples from the book, in debates about the "making" and remaking of society through the educational processes to form the child as the future citizen. Pragmatism combined with different sets of ideas and authority relations as people talked of a school pedagogy that made the individual as an agent of change, science as mode of organizing daily life, and the constituting of communities through processes of problem solving.

But the traveling of Dewey's pragmatism was not only as an icon of progressive change. It became the target of intellectuals, churchmen, and politicians who had different cultural theses about the moral underpinnings of the school and society. The international field of debate in which John Dewey and pragmatism enter makes possible the study of changes in the cultural patterns of schooling in regulating the constructions of the "modern" child and societies.

If I step back for a moment to ask about the cultural thesis of a *modern self* embodied in Dewey's pragmatism, it presupposes radical political theses about the individual as a purposeful agent of change in a world filled with contingency. The stability and order of the world are no longer viewed as the responsibility of the aristocracy, intellectual strata, and/or the ecclesiastical order. Change and stability are democratized in the sense that individual purpose and agency are to give stability and consensus to a world constituted as in constant change and constant uncertainty about the future.

This general schema about liberty and freedom, to use a liberal theorization of individuality, is not something found in the "nature" of the individual. It is something that is socially produced. Agency is not about total freedom but about the inscription of rules and standards of reason through which individuals live in social settings. Theories of the child and family, for example, fabricate an individuality that uses reason and rationality; and in some of the stories of modernities, these theories are meant to navigate the intricate webs of daily life and provide for a more progressive future. This is where science becomes important as a generalized mode of living that entails rationalizing and ordering participation and intervention in one's personal as well as social life. The "modern" self, in this brief schematic, is a particular historic invention of one who plans and orders actions in a rational way to bring about progress in a world of uncertainty.

Once I link the modern self to concepts of agency, science, and progress, two things come to mind about the historical quality of these concepts. First, the agency of the "modern" self is not something that the individual is born with but something that is "made." When one reads the early writings of the American or French Revolutions and subsequent writers forming educational systems, for example, schooling serves to fabricate the child who participates and acts as a reasonable citizen. This administration of the individual was seen as a prerequisite for the future of the Republic and later its notions of democracy. But agency also has its followers in the reforms of Brazil, Japan, China, and Turkey, though with different enclosures and internments of the individual. Second, science and agency are not concepts of logic or that of positivism alone. Science, as I use it, is a particular historical quality of the "mind" about planning and ordering the future in the present that forms in cultural practices. Agency and science are not one singular set of concepts but multiple ones that need interrogation.

This is where pragmatism and the writings of John Dewey come into play in this book. Pragmatism was a designing project of the individual who embraces the norms and values of agency, science, and progress. Dewey's pragmatism traveled as these notions were ordered, connected, and disconnected to talk about who the "new" child is and should be. In diverse nations that span Asia, North and South America, and Europe, intellectuals, government officials, and teachers sought a New Education. The term that circulated in the first decades of the twentieth century, to reform society through generating pedagogical principles for reflection, action, and participation. The book focuses on twelve countries to explore how Dewey enters into debates about the remaking of society through remaking the child. Dewey's pragmatism in the first decades of the twentieth century is reassembled in the salvation narratives of the Mexican Revolution, Chinese reforms to replace the Confucius hierarchies, Turkish modernization of its peasantry, the joining of Republicanism and Calvinism in Switzerland, and in a Pan-Slavic nationalism in Yugoslavia; and he performs with others as antihero foreigners who violate the *geist* of the nation (such as in Germany and Brazil), and as part of American imperialism by Mexican leftists and Maoists in China later in the century.

With these different "deweys" it is more appropriate to speak of plural *selves* and multiple modernities.

Historically studying Dewey as embodying a particular cultural thesis is to think of him as a *conceptual personae rather than as a particular icon of progressive schooling or as a creative person.*[2] Dewey enunciates particular solutions and plans for action in social, cultural, and educational arenas that go beyond its philosophical ideas. John Dewey was the international salesman for American pragmatism at a time when mass schooling was institutionalized in a diverse cultural and political field. Embodied in Dewey's writing is a particular set of concepts and ways of reasoning about the world and the self that is not merely that of Dewey. At one layer his writing brings together and articulates a particular system of reasoning that embodies cultural, social, and political conditions that gave that thought a plausibility and intelligibility. At the same time, pragmatism enters different cultural battles about the spiritual, moral, and social qualities of the child as it moves within the U.S. and in international field about educational reform.

In a sense, this book plays with Dewey's own title, *How We Think* (1910). The book asks, "How is Dewey possible as a traveling salesman and what was the competition?" But as significant, this gives focus to "What is the field in which he is put-in-play and in other instances, taken out of play?" "What is the grid that makes Dewey intelligible, accessible, sensible and relevant in places as different as Yugoslavia, China, Turkey and Mexico? And yet in other places and times, how is it that Dewey is targeted as the mortal antagonist that if allowed into the gates of the nation will destroy the consensus and stability?" Examining the travels of Dewey as a conceptual personae makes it possible to understand the cultural politics of the enclosures and internments of reflection and participation that move into the present, albeit in the names of learning, child development, and other benevolences about progress and well-being of the child.

Oddly, mass schooling is virtually ignored in contemporary scholarship about modernity and contemporary discussions of globalization, appearing only perhaps as footnotes to institutional developments or as an economic reductionism. Modernization literatures look at the more exotic of political and social–economic changes with little or no attention to the cultural theses in which its conditions of possibility were produced. The case studies in this book speak to that omission. The studies enter into a conversation with a major theme of sociology and political science: the twin and mutually related processes of globalization and localization during the twentieth century. These processes are recognized by focusing on different cultural theses about the governing of the self in the governing of the nation. The discussions of this book at the same time contribute comparatively to contemporary scholarship about John Dewey and pragmatism. And finally, as the Belgium study in this book argues persuasively, the methodological issues for studying the flow of ideas and institutions as material practices are complex and filled with theoretical issues that are not resolved and need constant scrutiny.

Reading Historically about How We Think

I begin with a biographical note about Dewey to juxtapose it with then another historical reading. Here, one would say that Dewey was born in 1859, the year that Darwin published *Origin of the Species*, and dying almost a century later, Dewey wrote across the domains of political theory, philosophy, psychology, and education, but also as a public intellectual. He was a quintessential American thinker of his time, a time in which the nation he "represented" rose to international significance. His first long work on education "the School and Society" was published in 1900 in Britain and the first translation was probably in 1904 in Czech.[3] Dewey lectured about pragmatism and/or his books appeared in such diverse places as Australia, Brazil, Britain, China, Columbia, Germany, Japan, Finland, Mexico, Portugal, Serbia, the former Soviet Union, Spain, Sweden, Switzerland, and Turkey.

Renewed interest in John Dewey talks about him internationally as instructive for the thinking about contemporary social life, public philosophy, and educational challenges. A recent book review likened the influential contemporary German social philosopher Jürgen Habermas to Dewey. Dewey is spoken of as "a professor of philosophy and social theory on one hand and a political controversialist" whose public presence is still felt. Dewey, the argument continues, "replaces transcendental guarantees about what is true, good, beautiful and 'the real' with methods of action about how to make the world more rational and just."[4] The progressive individual is associated with new versions of modernity that have prefixes affixed to its noun quality—post, neo, advanced.

The notions of biography and author in the above history are difficult to put aside. When first talking about this project with colleagues, the initial response was to compare what Dewey actually said to how that "saying" was interpreted, influenced, or borrowed in different national contexts. Biography and intellectual history was a "natural" starting point to explain variations or the corruptions of ideas as one moves from place to place or in different time frames.[5] This thinking about the author places Dewey as the originator of thought to assess others' faithfulness or abuse of the ideas. The task of finding truth is to isolate the true analytical qualities of thought. Central concepts are identified as entities that exist independent of their textual use and the cultural practices that produce them. A major journal of the American Educational Research Association, *The Educational Researcher*, continually runs essays and commentaries about who has been the most faithful to Dewey in an ostensible search to find the final, and correct reading.

The "telling the truth" about the author's real meaning is deeply embodied in European modernity and in the twin notions of doubt and the transcendental in literacy and reflection. The legacy of the true analytical "eye" is itself a tradition of the modern that ignores a different tradition of modernity. That tradition is to explore how words and their concepts are produced historically as objects of reflection and action. The analytic reading that looks solely at the logical structure of a concept denudes "thought" of the complexity that Bakhtin expressed when he said: "Language is not a neutral medium that passes freely and easily into the private property of the speaker's intentions; it is populated—over-populated [-] with the intentions of others. Expropriating it, forcing it to submit to one's own intentions and accents, is a difficult and complicated process."[6]

While I do not want to detract from the particular contributions of Dewey to philosophy and education—his position as an author—I do want to historicize that authorship to consider the issues of governing the "self" presupposed in the productions of modernity and the "modern" school.[7]

The chapter proceeds by first discussing two concepts through which to think of Dewey as a conceptual personae: notions of *indigenous foreigner* and *traveling libraries*. I then approach the reading of Dewey as a problem of historicizing ideas as cultural practices. The third section gives attention to modernity as sets of cultural theses through which modern selves are given intelligibility. This entails three concerns: the invention of human agency, the idea that biography is planned through calculating the rules and standards of reason and participation, and the inscription of science as a method of living that connects individuals with collective visions of progress and the nation. The fourth section focuses on agency as a concept that embodies the expertise of pedagogical and social sciences that renders reflection and participation as administrable. The final section focuses on science as not merely a way of ordering and rationalizing one's environment but embodying different notions of the sublimes that travel with agency and collective visions of different modernities.

Historicizing Ideas: Dewey's Pragmatism as a Conceptual Personae

Two intellectual "tools" are used to read "Dewey" historically as a conceptual personae. One is the phrase *the indigenous foreigner*. The other is *traveling libraries*. These two notions provide a way to explore the field of cultural theses about *modern selves* in which pragmatism traveled and connected/disconnected with other systems of reason.

First, I use the ironic phrase *indigenous foreigner* to give attention to how particular ideas as modes of living are brought into new contexts in which the "foreignness" of the ideas are seen as indigenous or ahistorical and "natural" to that situation in which they are positioned. John Dewey is one such foreigner. He travels to other lands where his foreign status is often forgotten as he "becomes" a native son. Today such indigenous foreigners might also be Michel Foucault or Paulo Friere. They, as is Dewey, are reassembled in different places as local reformers seek to change the principles for creating belonging and collective homes. The particular Deweyian concepts of action or child-centered education, for example, are "made" in America but this foreignness does not seem relevant as the concepts move into global discourses about a new education that crossed ideological lines. It was allied with eugenics in some places, in others with authoritarian regimes and anarchism, and in other with revolutionary salvation narratives in creating a strong national culture. Dewey's notions of the child, for example, are invoked as *indigenous* to bless social reforms that link the individual to society and the nation in the name of progress and change.

The *indigenous foreigner* is a notion to focus historically on the flow of systems of reason such as pragmatism as it reassembled, connected, and disconnected to articulate multiple intentions and cultural theses.[8] For example, Dewey is taken up in Sweden today as part of a "restorative democracy" to respond to what is perceived as threats to the welfare state. But the Swedish Dewey does not reflect the American belief in its national destiny embodied in Dewey's writing, but as a folk hero of the images of the benevolent patriarchical welfare state. In Turkey, Dewey appears as a modernizing figure to create a strong national culture among its peasants and in Columbia to stabilize the hierarchy of class divisions in the early reforms and later as a benevolent figure in the reconstruction of Columbian society. As today's educational reforms travel across the globe, John Dewey still functions as an indigenous foreigner, cojoined with the Russian psychologist Lev Vygotsky in the new progressive pedagogy.[9] Dewey wrote in the context of turn of the twentieth-century American social gospel movement to bring Christian ethics into social policy. The other, vygotsky sought to find a psychology that articulated the moral commitments of the new Soviet regime. Both are dead now. Their "history" emptied, to borrow from Walter Benjamin, as the two foreigners are joined at the hip as universal heroes of the "new" reform pedagogies in South Africa, Spain, the Scandinavian countries, and the United States among others.[10]

The indigenous foreigner involves a second concept, that of *traveling libraries*.[11] The notion of traveling libraries provides a way to think about what appears on the surface as the anomaly of pragmatism traveling with ideological positions with what seems as strange bedfellows. If we follow the paths of Dewey's writings, they circulate and reassemble with other American pragmatists such as William James; and they join in with conversations of French and German pedagogues that, in turn, are joined in different amalgamation of "local" authors in the Balkans, Belgium, Turkey, and Mexico.

The metaphor of *traveling libraries* gives focus to the different sets of assemblages, flows, and networks through which intelligibility is given to the changes. Dewey, for example, functions in many contexts as a metonym for modern pedagogy, the New Education, and the cosmopolitan child and society. The ideas and concepts of Dewey, for example, are assembled with the Swiss pedagogue Claparède and the Belgian Decroly in South America for national reformers to announce "the New Education." While Claparède, Decroly, and Dewey traveled together in many places, that does not tell us of the different amalgamations of pedagogical projects as they were not just one traveling show. Decroly served as a translator of Dewey in a Belgian missionary, evangelistic, and propagandistic pedagogical discourse. Pedagogy was to keep Christian doctrine as a safeguard of the order of progress

Some Fellow Travelers

Turkey

The photograph is a montage of people and programs in which Dewey traveled. Hasan Ali Yücel, Minister of Education in Turkey from 1938 to 1946, the students of Village Institutes, and the graduates of teacher-training courses. The students of the Institutes worked together to build and maintain the schools. The Village Institutes project was launched during Yucel's ministry. Together with İsmail Hakkı Tonguç, he was the second architect of the Village Institutes for teacher training of peasant youth. The teacher-training courses preceded the Village Institutes, to meet the "urgent" need for teachers to work in the village. This photo also illustrates the army officers who had completed their six-month course in teacher training and were about to leave for the village to begin their teaching careers.

Photograph used with permission from "The Village Institutes Album" of the Turkish Ministry of Culture.

Hasan Ali Yücel [1897–1961]

Minister of Turkish National Education from 1938 to 1946. The Village Institutes Project was launched by his ministry. A graduate of the University of Istanbul with a degree in philosophy, he taught philosophy, literature, and Turkish at high schools in Turkey. Serving as a school inspector and director general of secondary education at the Ministry of Education, Yücel later was the president of Gazi University, Institute of Education in Ankara. He contributed articles to newspapers published poetry, and wrote in many fields, such as logic, education, and citizenship. The reforms of the era in which Yücel worked as the Minister of Education ranged from the establishment of a translation community aiming to render world classics into Turkish to the reorganization of higher education institutions, as well as the creation of the Village Institutes.

Photograph used with permission from "The Village Institutes Album" of the Turkish Ministry of Culture.

İsmail Hakki Tonguç [1897–1960]

Founder of Turkish Village Institutes for teacher training of peasants. To complete the Village Institute's Program, the Turkish government sent him to Germany to study the methods of teaching drawing, handwork, and sports. He assembled together the works of John Dewey and Georg Kerschensteiner in forming the curriculum as well as different Turkish movements that he brought into the designing of education.

Photograph used with permission from "The Village Institutes Album" of the Turkish Ministry of Culture.

Sweden

Alva Myrdahl [1902–1986]

Important ideologist in the development of the Swedish School and welfare system and one of the pioneers of a Swedish preschool. She wrote about the threats and anxieties that small children encounter in the emerging industrial world and she especially focused on the preschool and other welfare institutions as places where children would be saved from those threats. With her husband, Gunnar Myrdahl, she wrote about Dewey and also wrote about the need to improve society through improving human beings. She was the co-recipient with Alfonso García Robels of Mexico of the Nobel Prize for Peace in 1982.

Photo reference: Photographer, Unknown, 1937. From the Swedish Labor Movement Archives and Library.

Ernst Wigforss [1881–1977]

Swedish secretary of state for the Treasury and regarded as the most important Social Democratic ideologist in the construction of the Swedish welfare state after World War II. Dewey's ideas were related to William James in a Marxist and then later a socialist context concerned with adaptability and plasticity of human nature as they interact with their environment.

Photo reference: Karl Sandels Photo Agency, 1935. From the Swedish Labor Movement Archives and Library.

Portugal

Adolfo Lima [1874–1943]

Adolfo Lima (1874–1943) was the central pedagogue in the reception, structuring, and relaunching of the so-called New Education in Portugal. Furthermore, he founded and ran a pedagogical reflection journal, published a methodology and didactics manual for teachers, and, in addition a host of essays about education, even published a pedagogical encyclopedia. Lima did much more than merely disseminate the work of John Dewey Portugal from the mid 1920s onwards. Dewey functioned as a reference point that enabled Lima both to structure his entire educational discourse and to put forward proposals to transform Portuguese schools, which were immediately deemed credible by the different school authorities.

China

Jiangyan Hu Shih [1891–1962]

Founder of linguistic reform in China. He studied under John Dewey at Columbia University, becoming a life-long advocate of pragmatic education reforms as a piece of gradual cultural reforms in China in its first mass movement in modern Chinese history (1919). The mass movement, The May Fourth Movement, began a patriotic outburst of new urban intellectuals against foreign imperialists and warlords. The movement split into leftist and liberal wings. The latter advocated gradual cultural reform as exemplified by Hu Shih who interpreted the pragmatism of John Dewey, while leftists introduced Marxism and advocated political action. The movement also popularized vernacular literature, promoted political participation by women, and educational reforms.

Used with permission from the Hu Shih Memorial Acedemia Sinica.

Brazil

Anísio Teixeira [1902–1986]

Brazilian educator, translator of Dewey's works into Portuguese. He led educational reforms "inside" the state provoking opposition from Catholic Church and from other elements of the state.

Photos taken from the Anísio Teixeira Virtual Library www.prossiga.br/anisioteixeira, Brazilian site developed and maintained since 1998 by MEC/CNPq/Prossiga and the Anísio Teixeira Foundation: AT3.jpg—Anísio Teixeira Family Archive, ca. 1940.

Anísio Teixeira

Professor Anisio Teixeira receives the Teachers College medal, 1963.

Photos taken from the Anísio Teixeira Virtual Library www.prossiga.br/anisioteixeira, Brazilian site developed and maintained since 1998 by MEC/CNPq/Prossiga and the Anísio Teixeira Foundation: AT4.jpg—FGV-Getúlio Vargas Foundation, Columbia University, NewYork—United States.

through ordering children's lives.[12] Dewey was reassembled in Columbia through Decroly and "local" authors in a reactionary and conservative pedagogy;[13] and Dewey and Decroly were placed in the company of the German Kerschensteiner and Swiss Claparède as the philosopher of a social redemption that Yugoslavian pedagogic work would center on the child's activity.[14]

The indigenous foreigner and *traveling libraries* are intellectual tools to historically inquire about cultural theses through which the self is construed and constructed. The two concepts draw attention to global flows of ideas and authority relations in particular situations as new intentions are opened, and new opportunities and interactions are invoked that are historically presupposed in discussions of modernity. At the same time, the concepts provide a way to consider globalization and localization as related and overlapping processes rather than different and distinct. The different photographs that follow are to think about the different assemblages and connections in which Dewey participates in different traveling libraries. I have not included a photo of Dewey in this collage to emphasize the flows of ideas, their networks, and conditions and not to reinsert Dewey as an icon of change.

Some Observations about Modernities and the Governing of Thought and Reason

I turn attention now to a particular set of principles that circulate in the cultural theses about the modern self in the cases studies of this book. The cultural theses are given intelligibility through a grid that I explore through three historical distinctions that epistemologically function to order action and participation. They entail (a) the invention of individual as an agent of change. Human agency introduces the idea of the individual that has responsibility for person and collective progress; (b) the regulation of time in the planning of agency. The modern "self" is a planned biography that exists in a serial or rational flow of calculated time. The pedagogical notions of child development and Dewey's concept of action embody such a notion of a planned life in the sequence of regulated time. And (c) the inscription of science as a method of living. Imaginaries of science are fabricated as a mode of living and planning for one's biography and progress.

The three distinctions are treated historically as different and overlapping cultural qualities through which modern selves are fabricated.

They are posed as historical markers and not necessarily the words of the protagonists and antagonists of the historical narratives told in subsequent chapters. I do not think that, Dewey used the term agency, but as I suggest later its notions if not the word was embedded in pragmatism. Nor are these notions stable and fixed entities to measure along a continuum of modernities. Agency, for example, is not merely the effecting of one's self realization according to some universal notion of participation and empowerment. Agency is produced within a historical set of conditions for acting that include, among others, who is defined as the agent to "act" and who is not capable of such action; rules and standards for participation, and the relation of the actor's agency to the collective belonging of "society" and/or the nation. Science, as well, is not one universal set of rules but rules and standards worked into sets of values that form historically specific or "local" principles of reflection and participation. In Belgian, Brazilian, and Turkish pedagogy, the child who uses "science" is not the same across these different spaces.

The different themes are historical devises to think about the multiple systems of reason through which solutions and plans for action are inscribed in traveling libraries discussed in the book's chapters. I deploy these categories, therefore, not as normative or definitional. They perform as a strategy to think about the dynamic and diverse cultural practices. The sections that follow discuss the insertion of a new expertise of knowledge about society and individuality and the sublimes that travel with the different notions of agency and science as a method of ordering one's life.

Inventing Agency and the Modern "Self"

The "modern" notion of agency is part of the sacredness of contemporary social theory and political doctrines. Not to evoke the image of the actor who has agency is seen as offering a deterministic world that has no option for change. But agency as a subject of social theory has its irony. The agency is not there to be recouped but is made.

If I pursue the notion of agency with Foucault's "What is the enlightenment? *Was ist Aufklärung?*" he suggests that the attitude of the Enlightenment brings a particular individual who is both an object and subject of reflection.[15] The individual embodies a new episteme in which there is no longer a transcendental or universal structure for all knowledge but a conditional world with notions of continuous change. The external world seemed destabilized and the future no longer moves in a steady line from the past through the present to the future. Individuality is formed and expressed in a world of uncertainty, a world that pragmatism articulates about the self as "from the contingence that has made us what we are, the possibility of no longer being, doing, or thinking what we are, do or think."[16] The future of incessant change and innovation now constitutes the new certainties of the world and its individuality.

The construction of human agency is inscribed in individuality. The idea of agency as an *a priori* historical actor is a radical shift in moving the locus of truth from a divine subject to the human creative subject. One's good works are no longer to prepare for the afterlife but for the betterment of life on earth. Reason is the mechanism through which agency is effected! Reflection and participation provide order and stability in a world of an unfolding and cumulative growth or progress.

The focus on agency introduces a new range of thought about the acting subject. But the systems of reason to order agency was not something of pure logic, nor was it natural to the workings of "thought." Reason is made into something calculable and administrable. John Meyer argues, for example, theories of agency entailed the progressive discovery of human personality in the eighteenth and nineteenth centuries; that each person carries a whole system of motives and perceptions that reflect different biological and social forces through which the individual self is integrated and constituted as the moral subject of one's own actions.[17] The ordering procedures of rational knowledge are to govern the direction and possibilities of the future.

The calculation and administration of action and actors/agency were central to the international spread of mass education and the construction of the modern nation in the late nineteenth and early twentieth centuries.[18] Dewey's pragmatism fit these historical conditions. Pragmatism was a device to intervene in childhood with the intent of ultimately influencing what society would be.[19] Dewey spoke of that society through the norms and values of American cultural imaginaries of the citizen in a democracy. Action was a central concept for the planning of one's life as a series, for example, of problem solving that ordered experience and designed the future. Agency was expressed in the idea of the child's reflectivity and thought as intelligent action. "[T]hinking enables us to direct our activities with foresight and to plan according to ends-in-view, or purposes of which we are aware. It enables us to act in deliberate and intentional fashion to attain future objects or to come into command of what is now distant and lacking."[20]

The governing of the child was a response to the uncertainty of modernity but also embodies a cultural strategy that expressed fears about maintaining the qualities of a civilized population. Artificially intervening in the individual development of the child was required, for if not controlled properly, the individual character and dispositions could be potentially dangerous to the moral and spiritual future of the nation. "The existence of scientific method protects us also from a danger that attends the operation of men of unusual power; dangers of slavish imitation partisanship, and such jealous devotion to them and their work as to get in the way of further progress."[21] The systematic training in "thinking" was to prevent "the evil of the wrong kind of development is [as] . . . the power of thought . . . frees us from servile subjection to instinct, appetite, and routines."[22]

Agency is a particular moral and political outlook embodied in pragmatism. For Dewey, the connection of consciousness and the world through action was the essence of democracy as a mode of living. Dewey urged, for example, to stop "thinking of democracy as something institutional and external" and to see it as "a way of personal life," to realize that "democracy is a moral ideal and so far as it becomes a fact is a moral fact." Dewey's pragmatism emphasizes the individual who participates in communities to act on an environment of continual processes of change. The new system of reason was a philosophy that "could examine how change served specific purposes, how individual intelligences shaped things, how scientific administration might beget increments of justice and happiness."[23]

Linking of ideas of agency and progress in a contingent world in Dewey's pragmatism has different configurations in European thought. One can compare, for example, debates between science and religion in the nineteenth century or the differences between Marxist, phenomenological, and behavioral notions of agency to recognize that there are different inscriptions to the agents of change and the constitution of agency. The most prominent approach in modern social theory is related to the philosophy of consciousness and theories of action. In this scenario, the individual is a purposeful actor who produces change through one's actions—sometimes willfully, sometimes with unintended consequences.

Dewey's anthropological psychology focused on providing the general principles or dispositions of the individual in forming actions. The individual was a social self formed through the contexts of action. The processes of mediation were through interactions and communication patterns. The ordering and administrating of the processes of *action* fell into a discourse of social psychology.[24] The new theories of social psychology focused on language and systems of communication and interaction in education as "habits of thinking" that "create attitudes favorable to effective thought . . . ."[25]

Dewey functions as both an author and as a conceptual personae. As an author, Dewey brings together multiple different ways of thinking that circulated at the turn of the century in a manner that is innovative. At the same time, that writing embodies and expresses its conditions of possibility. There is a salvation narrative in Dewey's writing that links the agentive individual to a particular grid of social, cultural, and political relations. Its model of democracy contained the possibilities for social progress and individual happiness that were expressed within as attitude of American Enlightenment thought.[26] The pragmatic outlook embodied an American Exceptionalism that brought Calvinist reform notions of salvation into the vision of the nation and its Chosen People. The pragmatism affirmed the progress in America that revisioned the myth of America as a cultural frontier. The school was a site where experience reinscribed the life of the pioneer as "literally instruments of adjustment and the test of consequences."[27] Pragmatism was of a new frontier mentality of an "American civilization" as a "process of continual expansion and reconstruction." Dewey's individual was a "Pioneer American" that opened the universe "in which uncertainty, choice, hypotheses, novelties and possibilities are naturalized that come from experience of a pioneer America."[28]

From the above, it is possible to think of pragmatism as not only about "thought" but as a cultural thesis about a mode of life. It formulates elite ideas of the Enlightenment into a project of social administration of everyday life. A colleague at Teachers College Columbia University, Arthur Childs said that Dewey's pragmatism was "a system in which ordinary people are to reshape the conditions that mold his own experience within the context of their own on-going activities, all necessary institutions, and regulative principles and standards."[29] The school was to make the individual as a central actor of change and progress through providing the dispositions and style of living that enabled human agency.

It is also possible to think of Dewey as conceptually giving intelligibility to particular solutions and plans for action for governing the cultural changes that were felt in different geographical contexts. America's pragmatic approach was an exemplar to others as extracting the progressive values and rationalities of governing without valorizing its nationness and economic principles. The German sociologist Max Weber, for example, saw the making of a new society in the American experiment of democracy and pragmatism.[30] The Italian Marxist, Antonio Gramsci viewed American pragmatism as resonating with international changes about a way of living, *being*, and acting.[31] That pragmatic way of life was separated from American Fordism as a model by which Gramsci could search for exemplars for a more socially progressive society.

The conditions that pragmatism presupposed were part of a progressive thought that circulated the Atlantic. It embodied notions of People's School or Folk Schools, for example, that mixed cultural nationalism with cooperative living in Scandinavia.[32] If we trace the reassembly of pragmatism in Sweden, the New Education was tied to the formation of the modern welfare state.[33] At one level, the subject was to be an agent of his/her own destiny. Learning by doing was a core principle. That destiny, however, was part of the collective committment through which the individual participated in a society referred to by the Swedish state and social scientists as The People's Home. It was believed that for the egalitarian improvement of society ("a school for all"), the Swedish population must be reformed. This reformation of the self was possible as man and society had a plastic human nature. Schools were to empower through making self-governing future-oriented subjects that have pleasure in cooperative actions. Dewey was joined in the Swedish traveling library with the German Kershensteiner and the American Kilpatrick to focus on the natural interests of the child, individualization, creativity, and spontaneity.

Calvinism and Republican values were re-inscribed in pragmatism as the later traveled into the making of a strong Swiss identity. Swiss activity curriculum had children explore their interests and problem solve as a way of expressing agency. Tröhler[34] argues that the pedagogy was to develop a strong Swiss culture with individuals voluntarily joining the division of labor. Activity was for the social adjustment of the individual by incorporating Darwinian Protestant ideas into thinking about the organization of teaching. Claparède, the Swiss who "translated" Dewey, wrote about pedagogy as genetic and functional, with little attention to American pragmatism's social and political norms.

It is possible at this point to consider the pragmatic openness to change and agency as cast through different amalgamations of ideas, authority relations, and institutions. Agency is not a universal quality of life but a particular construction that is inscribed in the making of the "modern" self. That construction relates individuality and collective narratives. But to speak of agency is also to consider the historically formed grids through which it is given intelligibility.

The Regulation of Time in the Planning of Agency

Part of the assembly through which the modern self was constructed was a stabilizing and taming of change in a world viewed as contingent and uncertain. Pragmatism, as did other new mixtures of philosophy and social science, focused on a method of change that regulated, stabilized and tamed, the contingency and uncertainty of the future. The calculation of "reason" performed as a biographical "tool" that ordered the temporal development of the individual in a world seen as otherwise conditional, insecure, and ambiguous. Life was a dimension of regulated time and development that projected the future onto the present.

Imaginaries of science were a process of ordering daily life to provide stability and consenseus. Educational reformers across different geographical places discussed in this book used nationalizing procedures associated with science to undermine social and educational traditions of schooling as well as in making the child whose dispositions to act embody the cultural norms and values of the "modern"—but with different salvation themes. The concern with science as a method for action in American pragmatism, for example, was to shred the Old World's traditions as they prevented progress and salvation. Dewey wrote of William James in 1929 that he was "well within the bounds of moderation when he said that looking forward instead of backward, looking to what the world and life might become instead of to what they have been, is an alteration in the 'seat of authority'."[35] The future was to be without an authoritarian system of religious and civil institutions, and any fixed classes and ancient institutions. "The old culture is doomed for us because it was built upon an alliance of political and spiritual powers, an equilibrium of governing and leisure classes, which no longer exists."[36] Arthur Child, Dewey's colleague, quotes Dewey as "our life has no background of sanctified categories upon which we may fall back; we rely upon precedent as authority only to our own undoing . . . ."[37]

Time was inscribed as part of the rational knowledge that had practical uses in planning one's biography and contributing to social progress. The notion of time gave reference to the future as a regulatory norm for ordering the present. Pragmatism, for example, was viewed as a system of reason to humanize the creative power of science and "thereby to gain control of the future."[38] The controlling of the future is through developing a scientific habit of the mind!

The connection of the past, present, and the future as a regulatory norm has different configurations in Portugal. Adolfo Lima initiated the New Education in Portugal and "translated" Dewey to promote a Social School where the child learned to solve problems and develop a mode of living that recognized multiple paths for action.[39] That agency was to use the interest that was innate to the child but for social purposes. The development of the child was to connect the past and future of the child's development in order to provide for progress of the nation that was "emancipatory" and free. While Lima was an anarchist, the authoritarian regime produced in the middle decades of the century inscribed the pedagogy for its national purposes.

The emptying of the past in the pragmatism of school reforms subdued history as no longer central to how one lives. But did it? The idea of tradition and its "elimination" is a creation of modernity. By identifying tradition with dogma and ignorance, Enlightenment thinkers sought to justify their absorption with the new. The faith in individual agency in the planning or actions toward the future outside of history's time was both to subdue previous traditions seen as hindering progress and to install new traditions, as I argue later, that associated the moral order of prior rural communities in the conditions of urbanization. The emptying of history entailed the production of new moral values and images of the sublime.

Science as a Practice in Daily Living

Science is a central marker to differentiate what is modern and what is not. The European Enlightenment thinkers of the eighteenth century sought a world citizen who used reason and rationality as a way of living among diverse cultures. In the nineteenth-century Europe and North America, scientific methods embodied what many intellectuals and social planners viewed as the most potent force shaping the world. The same practices that brought the natural world under control were thought of as able to order the social world for progress. Science not only assumed the mantle of the modern and of progress, but also as a way to destabilize existing hierarchies through reform.[40] Darwin's theory of evolution of the species, for example, was brought into a broader notion of social evolution in which social and individual worlds could be artificially modified in the name of progress and freedom.

Pragmatism embodied this faith in science that circulated in the latter part of the nineteenth century as a process in developing desired outcomes rather than in finding the rules concerning the unity of knowledge or Truth. The methods of science, it was thought, were to bring agency to individuals who effected change in their own lives and community. No longer searching for the unity of knowledge that would identify the providential rules of God on Earth, intellectuals placed their faith in the unity of the processes of science that enabled self-reflection and participation in a conditional world. The American universities, for example, began to define science as a disposition and method in a continual search for truth.[41] And this search for truth through science governed the principles of agency. For Dewey, "Command of scientific methods and systematized subject-matter liberates individuals; it enables them to see new problems, devise new procedures, and, in general, makes for diversification rather than for set uniformity."[42]

When I talk about science in the pragmatism of Dewey, it is important to focus on it as an embodied cultural practices for ordering daily life. Thus, I speak about science as a practice of acting that is abstracted from any particular institutional sense of it as what physicists or biologists actually do. The science of pragmatism is a fabrication to talk about and provide a calculated approach for ordering and classifying methods of governing individual actions. It provides a particular and socially framed system of standards and rules of reason and rationality that regulate individual life.

The faith in science had an almost millennialist belief of rational knowledge as positive force for action. Science was not only about the control of the external world. For Dewey and others, science was a mode of living that joined democratic processes and Christian reform notions of salvation in everyday living. Dewey saw no difference between a universalized notion of Christian values about the good works of the individual and the democracy that pragmatism embodies its notions of community, problem solving, experimentalism, and action. Dewey said that since "the future of our civilization depends upon the widening spread and deepening hold of the scientific habit of mind, the problem of problems in our education is therefore to discover how to mature and make effective this scientific habit."[43]

Pragmatism was to approximate rational inquiry of science in the conscious, daily search for individual and social betterment. Dewey, for example, looked to the experimental method of the natural sciences and the formulating of hypotheses to test theories as the best means of resolving conflicts about beliefs and revising our ideas in response to experience. Science was *plans of operation* that enable people to transform a given situation through resolute action in a world that is continually in the making. Agency is the application of science that gave intentionality and purpose to action. Acting "with an aim is all open with acting intelligently. To foresee a terminus of an act is to have a basis upon which to observe, to select, and to order objects and our own capacities"[44] "[S]cience signifies . . . the existence of systematic methods of inquiry, which, when they are brought to bear on a range of facts, enable us to understand them better and to control them more intelligently, less haphazardly and with less routine"[45]

This notion of science and planning involves a particular attitude that is assumed in modernity. That attitude is that the individual is both an object and subject of "thought." This "modern" individuality employs scientific thinking as a "tool" for developing abstract knowledge that enables one to reflect on the immediate without being tied to existing traditions and its common sense. "[T]here is no science without abstraction, and abstraction means fundamentally that certain occurrences are removed from the dimension of familiar experience into that of reflective or theoretical inquiry. . . . To be able to get away for the time being from entanglement in the urgencies and needs of immediate practical concerns is a condition of the origin of scientific treatment in any field."[46]

But the development of abstract knowledge and the removal of the traditions and ethos of one's immediate life was also a mode of living and acting. Science is a method of reflection and participation that recreates belonging and community that were seen as lost in urbanization, a matter that I take up later.

It is important to recognize that science as a method to organize life was the basis for the anthropological psychology brought into pedagogy. The pedagogical problems of "thought" as continuous problem solving in community and action had little to do with what scientists did, nor did it build on any actual studies of how science worked as a practice of inquiry.[47] The methods of science were a normative pedagogy for making the future citizen. They were practices to order thought and action related to future actions tied to a particular sublime that were embodied the notions of American democracy secularized Protestant notions of salvation and redemption. They were not a pedagogy of science, that is, how to learn about disciplinary norms, values, and cultural practices as systems of producing knowledge that circulated in fields such as physics or biology.

The Expertise of Science in Planning the Modern "Self"

The new expertise of science as a mechanism of change and the exemplar of what makes individuals as "modern" was central to *The New Educational Fellowship, Active Pedagogy* or *The New School* discussed in this volume. A rather heterogeneous school reform movement was initiated in the first decades of the last century that by 1932 represented in fifty three countries. Assembled in this movement are many of the people who formed the traveling libraries discussed in the subsequent chapters: Adolphe Ferriére, editor of the French journal of the movement, the Belgian Decroly, as well as the French Binet, Claparéd and Piaget, and the German Kerschensteiner.

The New Education Fellowship concerned transforming the pedagogical principles governing who the child is and should be. While not a monolithic movement, it promoted the scientific study of childhood and pedagogy to efficiently and effectively organize the school as a place of democratic living and progress, although these words had different frames of reference and authority relations in different places.[48] The sciences of pedagogy were not only a subject of inquiry into the processes and practices of teaching or a school subject for the child to explore the external world. Science as I talked about earlier, was a mode of life by which reflection and participation were regulated. Children's interest, diversity, creativity, and independence of thought were sites through which disposition toward acting were learned.[49]

The New Education was central to the reforms in different countries but with different cultural theses about agency as a mode of life.[50] Columbian Progressive Liberals in the 1930s, for example, reassembled Dewey in the construction of the New Education. These educational reformists saw pedagogy as a civilizing process that joined ecclesiastical authority with scientific reason as the child's way of living. Science as a process of living was bound to the unconscious modification of habits of action and in the process rejecting the idea of the autonomy of will power as separate from the faculty of the soul and center of spiritual life. The Mexican Revolution in the early decades of the twentieth century provided a different cultural thesis in the reformed school. The activity curriculum joined the sacred positions formerly devoted to Catholic religious emblems and traditions with notions drawn from the Enlightenment, rationalism, pragmatism, democracy, socialism, and republicanism. Turkish reforms of the same time selected particular pedagogical texts of Dewey for pedagogical reforms that would serve the secularization and modernization that the new regime had undertaken. The rationality of agency was subsumed in the name of the established good of the nation through redesigning the mode of living of the peasant child and family in rural schools. This new individuality would modernize the inner qualities of the child who would act according to the roles and practices envisioned by the enlightened elites of the nation. The dispositions of science and community meant to order daily practices were to place the individual within a hierarchy of social relations that placed the sovereignty of the nation as its pinnacle. In a different set of conditions and grid, Portuguese reforms inscribed science as a mode of living that integrated collective norms and values with a social-psychology of the child.

Pragmatism gave expression to different expertise in these cultural and social–political contexts. Where physics produced knowledge about the cosmos and supported technological innovation, the new American social and psychological sciences joined moral and secular values with the planning of the state, the organization of civil society, and the production of the citizen whose mode of life was to embody principles of reason and rationality. The revision of philosophy and the new disciplines of psychology and sociology, for example, directed attention to the practical issues of daily life. The new knowledge became empirical and instrumental, thought of as being useful for social and individual change. The progressivism of pragmatism "fitted" this program. Social psychology was concerned with the child as a self-monitoring and self-reflective organism that works for the future of "society." The sociology of the New Education in Portugal, in contrast, turned to Dewey to order and administer the social skills and sensitivities of the child for future use to society.

The professionalization of social and psychological knowledge was a part of what today might be called globalization. One can think of the European Enlightenment as multiple notions formulated and transmogrified through the world-wide travels and study by intellectuals. In the later nineteenth century, many founders of American social sciences and history traveled to Europe and studied in the new scientific institutions that were primarily located in Germany.

Dewey's entrance into this institutional setting was symbolic of the new professionalized sciences evolving in the borders of the Atlantic and beyond. At one level, Dewey was a professionalized intellectual who had a university position to pursue science as a paid career.

At a different level, the migrations were not only between the Atlantic, as is consistently identified in the following chapters. Japanese educators visited Johns Hopkins University, the first American university modeled after the German Humboldtian notion of a university as a site of the production of scientific knowledge, and then returned to Japan to promote notions of Dewey in education. Other pedagogues from South America, Asia, and Europe traveled to the United States for study trips or for graduate studies. For some, America was an exemplar to express the role of the individual in processes of modernization and the formation of conduct embodied in the new ideas of citizenship.

The new expertise was not only about the improvement of study of society and organizations through its empirical methods. The new expertise produced principles for governing the individual who participated in society.[51]

The new conceptual apparatus of sociology and psychology in pedagogy, for example, focused on the patterns of interactions of the individuals as domains of a moral community that was thought "lost" in urbanization and industrialization. The Chicago School of Community Sociology, which Dewey interacted with revisioned the distinction of the German sociologist Tönnies between *Gemeinschaft* and *Gesellschaft*.[52] *Gemeinschaft* was the imagined rural and pastoral vision of community where neighbors prior to modernity come closest to nature and God. The urban world at the turn of the century brought forth a more abstract and impersonal set of relations that Tönnies called *Gesellschaft*, where the moral and spiritual commitments of the pastoral world was no longer binding on people. The social psychology in Dewey and his friend, George Herbert Mead can be thought as a revisioning of the imagined *Gemeinschaft*'s pastoral, rural community of a face-to-face community into the industrial modern conditions of *Gesellschaft*. The revisioning was through a social-psychology that theorized about the processes of intersubjective mediation for the self-realization of the individual.[53] Intimate face-to-face interactions were given a new design, for example, through such concepts of primary groups. The concept of primary group was to produce personal bonds through inscribing social norms of communications in a fabricated "community."

The sciences of education would fulfill the promise of the future by reconfiguring the values and norms through which immigrant families and their children ordered and planned their life in the city. The concepts of social psychology provide ways to think of the "self" and belonging through an abstract knowledge that joined everyday choices and intentions to social interactions of the "group." The connections between child, family, and community in social policy, health, social science, and schooling joined the metaphorical "American family" of the nation with the diverse ethic and immigrant families and the child in the construction of identity.[54]

Dewey was a traveling American salesman who preached about the new knowledge of human science and its usefulness in educational reforms to accompany industrial and cultural changes. He was an itinerate salesperson of a knowledge that finds what works in making a better individual for a better world.[55] Dewey's missions to China, Japan, the former Soviet Union, Mexico, and Turkey, for example, brought seemingly modern, cosmopolitan ideas into national debates about the making of society, the citizen, and education of the individual. Dewey's writings in the 1920s and 1930s were assembled in Australian and Turkish national narratives of culture, civility, and civilization.[56] Dewey, for example, championed the case of progressive educational and the new social sciences as the most effective instruments of human growth and social progress in a democratic society. This new kind of intellectual was associated with *being* modern.

The colonial contexts of Africa, not discussed in this volume, offer a "Dewey" in a complex library related to the social gospel movement brought by Protestant missionaries, Anglo-American foundations such as Phelps-Stoke Fund reports (as well as Carnegie and Rockefeller) and research conducted at Teachers College, Columbia University in Africa. Concepts of Dewey are transmogrified and coupled with those of individual advancement, community progress, and vocational and practical (agricultural) types of education directed to the African populations.[57] Dewey was first hesitant using Third World countries as laboratories, but became interested in bringing the new educational theories into colonial contexts as a way to promote a more enlightened citizen and nation.[58]

Disenchantments/Enchantments: Modernities and New Millennial Outlooks

Previously I spoke about the belief of science as a mode of life that replaced entrenched traditions, yet at the same time installing a new assembly of faith and science as the motors of progress. The notions of agency, science, and progress embodied both a loss of faith and a construction of faith. An enchanted world wasn't purged but reset as different frames of moral and redemptive values reworked and assembled with the cognitive operations that ordered reason, rationality, and progress. The modernities discussed in this volume embody different intersections between science and the sublime as cultural theses that order who one is and should be.

At first, the emphasis on science brings up the image of the disenchantment with the world. Science is given value as a method of acting whose neutral commitment to the pursuit of truth seems to stand outside of provincial values, local traditions, and ideological sides.

But this notion of disenchantment is not all it seems. Max Weber suggests that the science about people was never about a dispassionate thought. It brought to bear particular Puritan notions of salvation into a secular world to provide a sublime in which senses of awe, aesthetics, beauty, and fears traveled with the objects of reflection and action. The disenchantment was continually doubling with enchantments.

The relation of science to enchantments is itself a central theme in pragmatism as it embedded the individual in the forming of a moral and ethical social world. While post-Darwinist culture in the United States replaced scripture with nature, philosophy, and social science, religion was never left behind but turned toward scientific naturalism. The generalized faith in science associated with pragmatism and modernity, for example, revisioned redemptive themes of the Reformation through which the good works of individuals provide paths for salvation.[59] Science was a way of achieving a unity of the spiritual/moral. In William James's words, "democracy is a kind of religion," and for pragmatic reasons "we are bound not to admit its failure." Such "faiths and utopias are the noblest exercise of human reason," and from this was created an optimism in the future.[60] He thought that all that was wrong with us was that the Christian ideal of fraternity had not yet been achieved—society had not yet become pervasively democratic where its problems were solved by people and not solved by making proper connection with higher powers. Dewey's problem solving and experimentalism was a redemptive theme tied to how one thought, saw, talked, and acted in daily life.

The new cultural theses that overlapped enchantments and disenchantments became part of the struggles about schooling as inscribing modes of lives, particularly in countries where there was a strong relation between the state, institutional religion, and schools. When moving to South America, China, and Japan, for example, reformers evoke the need to shred old traditions in the search for a new society that embodied new sublimes.

The emptying of history in the modern "self" inscribed spiritual values. Dewey's travels in Yugoslavia in the 1920s and 1930s, for example, were to make Slavic modernity through ideas of a desacralized action and agency. Yet at the same time, the child was viewed as embodying divine or natural forces of a "Slavic Soul" that valorized the rural and village as the true source of the purpose of action and agency.[61] Chinese reforms in the 1920s transmogrified Dewey's philosophical and pedagogical notions to make a social and political space for reforming intellectuals in The May Fourth Movement.[62] The introduction of vernacular language, literary changes that valued the individual author, and child-centered education were to sanctify individual rights through one's location in a group, community, and nation. This sanctification of individual rights challenged the existing hierarchy of the Confucius traditions. The reformers sought to reinsert a different hierarchy through which a collective culture could be produced. That hierarchy had little to do with the liberal tradition that Dewey wrote about. A similar observation can be made of Japan as Dewey was placed in a curriculum to reorder the social hierarchy. Each student was placed within a lineal horizon of the family that reinscribed the relation of the child to the family and state as one continuous horizon, thus revealing the national character. Finally, Anísio Teixeira, considered the major interpreter and disseminator of Dewey in Brazil, sought to protect Dewey (and himself) from the charge of an anti-ethical rationalism and secularism as other Catholics reformers saw themselves as protecting society through inculcating Church values.[63]

Modernities, "Modern" Minds, and Schooling

The focus of this introduction is provide a way to think about the multiple ideas that flow in the making of the title of this book, *Inventing the Modern Self and John Dewey: Modernities and the Traveling of Pragmatism in Education.* The historical approach to Dewey as an indigenous foreigner and in a traveling library gives attention to "ideas" as cultural theses about a mode of living. I provided some "markers" through which to think about the construction of the "self" that traveled with and presupposed in the making of different modernities. Among these characteristics were the overlapping of a faith in a priori *actor's agency, the ordering and fashioning of a mode of living through principles of science, and the taming of change as a strategy of stabilizing and regulating conduct* in processes of change. The methods of science were to reform. These qualities of individuality overlap through methods of abstracting and "reflecting" on the present world. The "disenchantments" embodied new enchantments of attachment and belonging that today are embodied in discussions of modernity. The cultural practices in which agency, science, and the taming of change occurred were not variations of a single theme but of different configurations that related collective identities with individuality. The resultant cultural theses were assembled as traveling libraries that fashioned and shaped what constitute modernity and the "modern" selves. Dewey in this context functioned not as an original author but as a conceptual personae that connected and disconnected in different conditions to produce principles to order who the child is and should be in one of its major institutions through which the reform of society as "modern" was to occur.

The historical approach taken in this book of Dewey as an indigenous foreigner is to counter what Walter Benjamin has called an *emptying of history.*[64] That empty history has Dewey appearing as a logical system of thought or as "concepts" that have no social mooring in the interpretations and possibilities of action. The cultural and social history of this volume explores the multiple flows and amalgamation of practices (traveling libraries) through which "thought" is produced. Approaching the study of Dewey in this manner is to recognize that pragmatism is one of the multiple systems of "thoughts" about the principles that govern the relation of collective norms and values to individuality.

Approaching Dewey as an indigenous foreigner and part of traveling libraries relinquishes an old and, I think, fruitless debate as to whether Dewey's influence has been pervasive in theory but much less so in practice. This volume places Dewey with particular historical conditions and an international field of ideas, authority relations, and institutions through which modes of living were fabricated. The approach argues against treating modernity as an epochal concept of uniformities in social experiences and instead focuses on its nonuniform differences, distinctions, and divisions. Dewey is used in a strategy to explore school pedagogy as a major site in the production of the self in contemporary life. Historicizing the cultural practices that produce patterns of thought also provides a way of interpreting the relation of the global and the local, as well as the universal to the particular. This strategy does not center on Dewey's contributions, but rather inquires into the different cultural and social settings through which his ideas traveled and mutated to give intelligibility to different worlds called "modern."

Notes

1. I use the notion of *the long nineteenth century* to think of different historical patterns that move unevenly from the eighteenth century through the early decades of the twentieth century. I was told by an historian that the notion of "modern" appears in a German text in the sixth century, meaning of the present and recent times. While interesting, this current discussion is neither a history of evolution nor of origins.
2. I borrow this from Gilles Deleuze and Felix Guattari, *What is Philosophy?* trans. Hugh Tomlinson and Graham Burchell (New York: Columbia University Press, 1991/1994).
3. William Brickman, "John Dewey's foreign reputation as an educator," *School and Society* 70, no. 181 (1949): 257–264.
4. Alan Ryan, "The power of positive thinking," *The New York Review of Books*, January 16, 2003, 50, no. 1.
5. This is endemic in the literature. See, e.g., James Kloppenberg, "Pragmatism: An old name for some new ways of thinking?" in *The Revival of Pragmatism. New Essays on Social Thought, Law, and Culture*, ed. Morris Dickstein (Durham: Duke University Press, 1998) (83–127). For a reading that is closer to what I am talking about here, see Gert Biesta and Siebren Miedema, "Context and interaction. How to assess Dewey's influence on educational reform in Europe?" *Studies in Philosophy and Education* 19 (2000): 21–37.
6. Mikhail Bakhtin, *The Dialogic Imagination: Four Essays*, trans. Caryl Emerson and Michael Holquist (Austin, TX: University of Texas Press, 1981), 294.
7. While I have not foregone putting my name on this chapter or the book for my biographical reasons, I recognize it has multiple meanings and ironies.
8. For related work in comparative education, see, e.g., G. Steiner-Khamsi, ed. *The Global Politics of Educational Borrowing and Lending* (New York: Teachers College Press, 2004).
9. Thomas S. Popkewitz, "Dewey, Vygotsky, and the social administration of the individual: Constructivist pedagogy as systems of ideas in historical spaces," *American Educational Research Journal* 35, no. 4 (1998): 535–570.
10. Walter Benjamin, "Theses on the philosophy of history," in *Illuminations: Essays and reflections*, trans. Harry Zohn, edited and with an introduction by Hannah Arendt (New York: Schocken Books, 1955/1985), 253–264.
11. This phrase emerged at a meeting discussing the chapters in this book and I need to acknowledge Marc Depaepe's playfulness when introducing this term into our conversation.
12. See De Coster et al., chapter 4, this volume.
13. See Sáenz-Obregón, chapter 10, this volume.
14. See Sobe, Chapter 6, this volume
15. Micheal Foucault, "What is the enlightenment? Was ist Auflärlung?" in *The Foucault Reader*, ed. Paul Rabinow (New York: Pantheon Books, 1984), 32–51.
16. Michel Foucault, "What is the enlightenment?" 46.
17. John W. Meyer, "Myths of socialization and of personality," in *Reconstructing Individualism: Autonomy, Individuality, and the Self in Western Thought*, ed.

Thomas C. Heller, Morton Sosna, and David E. Wellbery (Stanford, CA: Stanford University Press, 1986), 208–221.

18. John W. Meyer, John Boli, George M. Thomas, and Franciso O. Ramirez, "World society and the nation-state," *American Journal of Sociology* 103, no. 1 (1997): 144–181.

19. Paul S. Boyer, *Urban Masses and Moral Order in America, 1820–1920* (Cambridge, MA: Harvard University Press, 1978).

20. John Dewey, *How We Think: A Restatement of the Relation of Reflective Thinking to the Educative Process* (Boston, MA: Houghton-Mifflin Co., 1933/1998), 16.

21. John Dewey, "American education and culture," in *Character and Events: Popular Essays in Social and Political Philosophy*, Vol. II, ed. Joseph Ratner (New York, H. Holt and Company, 1916/1929), 11.

22. Dewey, "American education and culture," 23.

23. Bruce Kuklick, *Churchmen and Philosophers: From Jonathan Edwards to John Dewey* (New Haven, CT: Yale University Press, 1985), 247–248.

24. See, e.g., Dorothy Ross, *The Origins of American Social Science* (New York: Cambridge University Press, 1991), 230–239.

25. P. Boyer, *Urban Masses and Moral Order in America, 1820–1920*, 73, 79.

26. The American Exceptionalism was never totally inclusionary; nor is that notion static. Dewey's pragmatism is part of a revisioning that begins to look to the future and its potential mode of living.

27. Robert A. Nisbet, *History of the Idea of Progress* (New York: Basic Books, 1979), 182.

28. Nisbet, *History of the Idea*, 11.

29. John L. Childs, *American Pragmatism and Education: An Interpretation and Criticism* (New York: Henry Holt and Co., 1956), 3–4.

30. Max Weber, *From Max Weber: Essays in Sociology*, trans. edited and with an introduction by H. H. Gerth and C. W. Mills (New York: Oxford University Press, 1946).

31. Antonio Gramsci, *Sections from the Prison Notebooks of Antonio Gramsci* (New York: International Publishers, 1971).

32. Daniel T. Rodgers, *Atlantic Crossing: Social Politics in a Progressive Age* (Cambridge, MA: Belknap Press of Harvard University Press, 1998).

33. See Olsson and Petersson, chapter 2, this volume.

34. See Daniel Tröhler, chapter 3, this volume.

35. John P. Diggins, *The Promise of Pragmatism: Modernism and the Crisis of Knowledge and Authority* (Chicago: The University of Chicago Press, 1994), 39.

36. John Dewey, "The schools and social preparedness," in *Character and Events*, 501–502.

37. P. Boyer, *Urban Masses and Moral Order in America*.

38. Steven C. Rockefeller, *John Dewey: Religious Faith and Democratic Humanism* (New York: Columbia University Press, 1991), 3.

39. See do Ó, chapter 5, this volume.

40. Shmuel N. Eisenstadt, "Multiple modernities," *Daedalus*, 129, no. 1 (2000): 1–29; Bjorn Wittrock, "Modernity: One, none, or many? European origins and modernity as a global condition," *Daedalus*, 129, no. 1 (2000), 31–60.

41. Julie A. Reuben, *The Making of the Modern University: Intellectual Transformations and the Marginalization of Morality* (Chicago: The University of Chicago Press, 1996).

42. John Dewey, *The Sources of a Science of Education* (New York: Horace Liveright, 1929).

43. John Dewey cited in John P. Diggins, *The Promise of Pragmatism: Modernism and the Crisis of Knowledge and Authority* (Chicago: The University of Chicago Press, 1994), 227.

44. Dewey, "American education and culture."

45. Reuben, *The Making of the Modern University*, 28–29.

46. Reuben, *The Making of the Modern University*, 16.

47. John Rudolph, *History of Education Quarterly* 45, no. 3 (2005).

48. The notion of democracy and freedom, however, has multiple and complex sets of meaning as evident in the discussions of the various case studies.

49. Edward A. Krug, *The Shaping of the American High School, 1920–1941*, Vol. 2 (Madison: University of Wisconsin Press, 1972).

50. The following discussion is drawn from the chapters of Sabiha Bilgi and Seckin Ozsoy (chapter 7), Rosa N. Buenfil Burgos (chapter 8), and Javier Sáenz-Obregón (chapter 10), this volume.

51. John Dewey, *The Child and the Curriculum, the School, and Society* (Chicago: Phoenix Books, 1902, 1900/1956).

52. I discuss this in Thomas S. Popkewitz, "The reason of reason: Cosmopolitanism and the governing of schooling," in *Dangerous Coagulations: The Uses of Foucault in the Study of Education*, ed. Bernadette Baker and Katharina E. Heyning (New York: Peter Lang, 2004), 189–224.

53. This is discussed in Thomas S. Popkewitz, Ulf Olsson, Kenneth Petersson (in press), "The Learning Society, the Unfinished Cosmopolitan, and Governing Education, Public Health and Crime Prevention at the Beginning of the Twenty-First Century," *Educational Philosophy and Theory*.

54. Pricilla Wald, *Constituting Americans: Cultural Anxiety and Narrative Form* (Durham, NC: Duke University, 1995).

55. This notion of traveling salesman and putting Dewey into this space of the new expertism emerged from a discussion with Mirian Warde and graduate students at the Pontifical University of São Paulo. The discussion directed my attention to Dewey in a broader cultural historical literature about professionalism in the United States that I speak about here.

56. See Bilgi and Ozsoy this volume.
Lesley Dunt, *Speaking Worlds: The Australian Educators and John Dewey, 1890–1940* (Melbourne: History Department, The University of Melbourne, 1993).

57. I appreciate Ghita Steiner-Khamsi's and Ana Isabel Madeira's comments when I asked them about Dewey and African education. I realized as I was writing this introduction that there seemed almost a total omission (on my part as well as in the historical discussions) of Africa in the movements of Dewey and pragmatism. Steinar-Khamsi and Madeira were generous with their time in responding to my emails and directing me to readings. See, e.g., Gita Steiner-Khamsi and Herbert Quist, "The politics of educational borrowing: Reopening the case of

Achimota in British Ghana," *Comparative Education Review* 44, no. 3 (2000): 272–299; Gita Steiner-Khamsi, "Re-framing educational borrowing as a policy strategy," *Internationalisierung/ Internationalisation. Semantik und Bildungssystem in vergleichender Perspektive/ Comparing Educational Systems and Semantics* (Frankfurt am Main: Peter Lang Verlag, 2002), 60–66. Ana Isabel Madeira, "Portuguese, French and British Discourses on colonial education: Church-state relations, school expansion, and missionary competition in Africa, 1890–1930," *Paedagogica Historica* 41 (2005), 1–2, pp. 33–62.

58. See Ronald Goodenow and Robert Cowen, "The American school of education and the Third World in the twentieth century: Teachers College and Africa, 1920–1950," *History of Education* XV, no. 4. (1986): 271–289.

59. Max Weber, *The Protestant Ethic and the Spirit of Capitalism*, trans. Talcott Parsons (New York: Charles Scribner and Sons, 1904–1905/1958).

60. Dewey in Kloppenberg, "Pragmatism: An old name for some new ways of thinking?" 96.

61. See Noah Sobe, chapter 6, this volume.

62. See Jie Qi, chapter 11, this volume.

63. Similar arguments about the loss of tradition and the needs for collective spiritual and moral values were used in American arguments against pragmatism, not recognizing or rejecting its secularization of Puritan notions of redemption and salvation.

64. Benjamin, *Illuminations: Essays and Reflections*.

II

European Spaces: The Northern and Southern Ties

2

Dewey as an Epistemic Figure in the Swedish Discourse on Governing the Self

Ulf Olsson and Kenneth Petersson

Introduction

We are witnessing today the ideas of the American philosopher and social theorist John Dewey being revisited and revived. In the Swedish context this renewed interest is manifested in rather a constructivist way. Dewey's thought on, for example, *freedom, democracy, community, and communication* during the early twentieth century is again being inscribed in contemporary narratives about society and citizens in relation to the future, that is, the constructivist project as shaping the current conditions for self-governance in the name of those concepts.

Dewey was, during all of his intellectual life, engaged and occupied with ideas about the future-oriented individual and his/her relationship with the questions of community, democracy, and education. Contemporary questions, taking up a great deal of space in the discourse concerning the remolding of the relationship between the civic subject and society are not essentially different from those Dewey brought into fashion in the early 1900s. In the early twentieth-century Swedish context Dewey's thought became linked to the system of governmental discourse on individuality and modes of living responding to rapid changes in Swedish society. In the current constructivist "paradigm," Dewey is emerging as part of a *traveling library* in a new system of governance concerned with questions and discussions of democracy relating to notions of pluralistic and multiple societies. In this way Dewey has reincarnated into a new figure and has been reestablished as an *indigenous foreigner* whose ideas are viewed in relation to and linked with current educational and political discourses. But rather than a reincarnation, what we are witnessing in current constructivist theories and practices is the emergence of a new Dewey, different from the "original" one. While the name of Dewey, as intellectual author still remains, contemporary political reforms "exist within an amalgamation of institutions, ideas, and technologies that is significantly different from those of the turn of the century."[1]

Dewey is becoming thought of in a different way, thus, as a paradigmatic case in the Swedish reform discourse, particularly in relation to the refashioning of the relationship between democracy, the individual self, and education. Dewey is reinstated, not as a figure of the past, but as a contemporary productive figure of support in the reconstruction of the present and in the rediscovery of learning and education as significant technologies in the project of reframing and remolding governance of Man and Society. In this respect, the use of Dewey's educational, philosophical, and political thought could be viewed as a technology in the service of shaping the future, different from the one molded during the end of the nineteenth and the beginning of the twentieth century.

We will endeavor to illustrate this from a genealogical point of view by looking more closely at the problem of producing future-oriented individuals in contemporary Swedish political discourses on democracy, the education and training of teachers, public health and criminal justice. Our main concern is to elucidate how the ideas of Dewey as an indigenous foreigner are assembled and connected with other discourses on the reconstruction of a national and collective sense of who and what the Swedish citizen is and ought to be.

We begin this chapter by looking more closely at how Dewey's ideas were brought into play in the earlier periods of the Swedish welfare state. This is followed by a more general description of what characterizes the Swedish discourse on the relationships between Man, Society, and the Future in the beginning of the twenty-first century. Next, we illustrate the way in which Dewey as an indigenous foreigner is present in the different contemporary institutional narratives we have chosen to explore. We end with a discussion and some conclusions that point to changes in the way government reasons in the name of the learning society.

Dewey as an Epistemic Source of the Development and Formation the Swedish Version of the Welfare State, Folkhemmet, The People's Home

Dewey's philosophical thought and theories on education were operating in the political narrative of the 1930s and 1940s and in the vision of an alteration of power relations in society and the creation of a welfare state, as translated from the Swedish, The People's Home. The foundation of the People's Home is fellowship and a spirit of togetherness. The good home is not acquainted with privileged or disadvantaged people; there are no favourites and no one is excluded. In the good home nobody looks down on anyone else, nobody tries to get advantages at someone else's expense, the strong do not oppress or the exploit the meek. In the good home equality, consideration, cooperation, and helpfulness prevail.[2]

Dewey's ideas were operating in this vision of the Future expressed by influential Social Democratic politicians and educationalists. In their use of Dewey's ideas, the education system became a technology for the ambition to alter society in a socialist direction. Oskar Olsson, a prominent Social Democratic popular educationalist, held that school must become a democratic community with the main task to create a democratic society.[3] Stating this, Olsson was making references to Dewey; "democratic ends demand democratic methods for their realization."[4]

We can turn to Ernst Wigforss (1881–1977), secretary of state for the treasury and regarded as the most important Social Democratic ideologist in the construction of the Swedish welfare state after World War II. In his memoirs he emphasizes the importance of Dewey's ideas for the development of his own philosophical and political thought.[5] He was introduced to the pragmatic ideas of William James and John Dewey in a meeting of a group of radical students, which he described as his spiritual home around 1905. Through Wigforss, Dewey was inscribed in a Marxist and then later a socialist context. According to Wigforss, there is a lot of pragmatic thought in Marxism, for instance, the conviction that a human being is first of all an acting being interacting with his/her environment. In his memoirs, Wigforss wrote that it was a relief to read Dewey because he emphasized the constantly changing nature of human beings.

The adaptability and plasticity of human nature, that during the last decades, with abundance of concrete details has been illustrated by ethnologists and social psychologists, was a natural element in Dewey's psychology.[6]

Dewey's criticism of the conception of the human being as a "ready-made and finished self" was well matched to Wigforss's narrative about the reformation of Man and Society. In the narrative of society and future advocated by Wigforss, plasticity and adaptability were necessary characteristics. It was thus easy for Wigforss to bring Dewey's conception of Man into play in the political project of Social Democracy. Dewey's ideas became embedded in the narrative of The People's Home, and were a crucial part of the library of thoughts constituting the intellectual framework not only for the development of the postwar Swedish education system, but for the formation of the Swedish welfare state as a whole.

Other influential Social Democrats inscribing Dewey's ideas in their vision of the Future were Alva and Gunnar Myrdal. Referring to Dewey, they held that "society must become improved by improved human beings."[7] They wrote almost exuberantly about American society and about Dewey's experimental schools. In their opinion, the United States had become, thanks to Dewey, the forerunner for modernity in general and for education in particular.

Like Sweden, which has been a laboratory for progress within the field of social policy, the United States of America has become a laboratory in the field of education.[8]

It could be suggested that Dewey's thoughts to a great extent became instrumental, to the process that resulted in what can be described as a paradigmatic shift in the Swedish welfare project. This becomes clear in the report by the Education Commission, appointed by the government in 1946.[9] The Commission proposed the introduction of a nine-year basic compulsory education. However, the idea of abolishing the old class-stratified education system and replace it with a common basic school for children from all social groups, "a school for all" was not new. We are saying that "the school for all" is not new as an idea as Dewey's ideas were known in Sweden before Dewey appeared as an indigenous foreigner. As early as 1883, Fridtjuv Berg (1956–1916), who was the chairman of the Swedish Association of Public Primary School Teachers and Secretary of state in two liberal governments was advocating the introduction of an education system where every single child, in cooperation with children from all social backgrounds, could flourish and grow according to their own aptitude and interests.[10] Berg embodied a thought that made possible the linking of Dewey as an indigenous foreigner into his own narratives about School, Society, and Future. Thus, we could say that Dewey was here already before he arrived. He was present, as a poser of problems, as a figure of thought and discursive practice, long before he entered the Swedish educational arena in the name of Dewey.

But the crucial question is what vision of society and the future was Berg talking about? According to Berg, a stratified school system, that separated children from wealthier families from those from poorer backgrounds, was a threat to the development of individuals as well as to the development of society. He expected the basic education system to bridge the chasm between social classes and thus to reduce social conflict in society.

The divided school system breeds a split in peoples' mind, smothers and restrains the only force that could unify people in society, namely a sense of fellow-feeling.[11]

According to Berg, the coexistence in a school common to all, and the shared upbringing of children would create a New School and thus a new Citizen, capable of fellow-feeling, not only with their social group, but also with other social groups and with society as a whole. Berg was inscribing ideas of how citizens should live in a liberal political system of thought with the intention to create a school that, in his opinion, would be a more democratic and liberal school. But in the final analysis, it was not about the education system. It was about society, about education as a technology for managing increasing social disorder. In Berg's narrative Dewey's ideas were inscribed in the library of thoughts used to fabricate and develop a citizen capable of sympathy with others but who would not question fundamental power relations in society. Berg's Future was of the future of yesterday and in his program education became a technology targeting on the self of the individual and on the development of dispositions and capacities supporting this "Future." In both Berg's and Wigforss's cases the individual has to be reconstructed, but in the Berg vision the individual was thought of as helping to preserve society, while in Wigforss's and the Social Democrats' vision of the future, society needs reconstructing.

According to the vision of the People's Home, the main task of the education system was to bring up a democratic citizen and to develop democracy as a political project. School had to "establish a democratic spirit in people's minds."[12] Education became an essential technology for supporting political changes leading to the development of empowered, self-governed, and future-oriented subjects and consequently for a "rehabilitation" of society in the name of democracy. Of crucial importance was the question of freedom versus authority and how to utilize education as a technology for promoting the interest of the individual as well as of the collective.

The task of the democratic school, thus, is to develop free human beings who have a desire for and take pleasure in cooperative action.[13]

The modern future-oriented individual of the Commission's vision of a new education system, was a free and democratic human being with a deep interest in and ability to cooperate with others. Dewey's thought about "education as a freeing of individual capacity in a progressive growth directed to social aims"[14] was operating in that narrative. The Education Commission deeply questioned the traditional teacher-centered and obedience-oriented school. It referred to progressive writers such as the German Kersechensteiner, the Austrian Köhler, and the American Kilpatricks. The progressive conception of Dewey's thought that was reassembled in Sweden was concerned with the natural interest of the child, the child in focus, individualization, creativity, and spontaneity. School was seen as a laboratory for concrete and reality-based teaching with "learning by doing" as the core principle.

The ideas of Dewey inscribed into the narrative after World War II were not the same ideas that were inscribed into the narratives of the beginning of the twentieth century. Berg's narrative embodied figures of thought that made it possible to connect Dewey to a Swedish discourse about the future of yesterday. Wigforrs was inscribing Dewey's thought in quite a different narrative, the discourse of the idea of the People's Home, an idea with the power to change society. The vision of the People's Home was built on an idea that it was possible to foresee and make plans about the future and that it was possible to fabricate autonomous citizens that were dependable and willing to play their part in a collective nation-building enterprise. The kind of society created in this discourse was an all-embracing society, which came to be known as "Folkhemmet"—The People's Home—where people's lives are put in order. School was seen as a crucial practice in the shaping of the constitution of the individual with the kinds of dispositions and capacities needed to fulfill the aims of this project.

Dewey's ideas were alive in the discussions about the Swedish school system during the 1950s, 1960s, and 1970s and possibly longer. Direct references to Dewey were more apparent in the earlier parts of the period than in the latter. During the 1970s, when the understanding of school as an apparatus in the service of democracy was overshadowed by the conception of a value-free school, with vocational training in focus, Dewey's position was no longer as eminent. However, his ideas were still present in the thought of various progressive pedagogues.[15]

The Futures of Today and Yesterday

During the postwar period the welfare state emerged as an open field for an all-encompassing field of practice. It was during this period that Dewey once again became an indigenous foreigner in the Swedish educational and political context. Educational thought at this time addressed society in its entirety, where society equaled the welfare state. There was, for example, no distinction made between the state and civil society. The Swedish historian Yvonne Hirdman described this vision of the state as an all-embracing power with the task of ordering people's lives.[16] This had an obvious impact on educational reasoning. Inconsistencies in and failures of "the system" would be given structural explanations.[17] The shaping and building of knowledge and understanding in the other areas of society that we cover in this chapter, such as criminal justice[18] and public health[19] followed a similar course.

In this framework of knowledge and politics it was impossible to conceive of the individual subject as the agent of his or her destiny. The shaping of the future was a collective affair, and there were in fact very few special personal demands made on the individual as long as s/he were prepared to conform to the common standards of welfare.[20]

The 1990s mark the appearance of new welfare politics and new regimes of knowledge. Pedagogy, which in Sweden is the term name for educational science as well as for the practice of teaching, underwent a considerable transformation. While the pedagogy of the 1970s saw in society the main key to an understanding of educational processes, that of the 1990s limited its concern to Education proper. So what was formerly considered as determined from "without" (society) has now become something that is internal to education and can be manipulated and dealt with in the course of the educational process. The subject that fits this agenda is created in the absence of "society" that is replaced as the organizing principle of educational thought. The individual subject is now the organizer of his or her destiny and future, and the task of education is to empower the subjects to put their own lives in order themselves.[21]

In the following section we give examples of the inscription of Dewey's thought in contemporary Swedish discourses about governance.

"The Future Is Here and Now—What Are We Waiting For!"[22]

There is always a kind of mysteriousness about things, which deal with the future, as if it might be something decreed by Fate and out of sight of our own activities. The future, however, is not unexpected, something we are unable to defend ourselves against. The future does not consist of a number of strange spaces, into which people are entering by force [. . .]. The future is not a space into which people have to enter by force or by absolute necessity. The new society cannot be designed independently of people existing here and now. The future is designed by people present in society at this very moment.[23]

During the end of the twentieth and the beginning of the twenty-first century the future, and democracy as a method of governing the future was problematized once again. In narratives about the society of the future and about who and what the future-oriented child and adult citizen are and should be, figures of thought on Dewey's model are coming alive once more. In this context Dewey's thought is inscribed into most political areas, not only that of education. This comes as no surprise given that contemporary educational thought is, in the name of lifelong learning, inscribed into different political and institutional areas that were not previously conceived of as educational.

The concept of lifelong learning crosses the boundary between political sectors. Educational policy, employment policy, the policies of industry and commerce, regional development policy, and social policy all have a common responsibility for lifelong and life-wide learning.[24]

The contemporary constructivist discourses about Man, in sectors such as education, public health, criminal justice, and the reformation of democracy, are organized through a fairly distinct epistemological language for shaping ideas of a new type of citizen or organization; one that is autonomous, responsible, learning, and problem solving.

The citizens have to be equipped with autonomy that enables them—in different kinds of mutual self-government and to the greatest possible extent—to order their lives for themselves.[25]

This notion of the citizen of the future, is based on two assumptions in the Swedish Democracy Report.[26] The first is that the life-projects of the citizens are developing in increasingly heterogeneous directions, which brings about risks, fragmentation, and moral crises. The individual life-project becomes estranged and the spirit of solidarity is lost. The second assumption is that this development needs counteracting by nurturing the receptiveness in the citizens to the ideals of democracy as applied to their own lives. According to the Committee Report, in order to realize the ideals of a society where the individuals are willing and able to participate and act in a spirit of mutual solidarity in a context of pluralism and fragmentation, a certain kind of autonomy has to be developed and nurtured. Education as a School of Democracy will thus be central in shaping and developing the autonomous individual and encourage him/her to take responsibility not only for him/her self but also for creating a common community narrative as part of his/her own life-project. Contemporary government reasoning adopts this extended and paradoxical route in an attempt to bridge the increasing and unforeseeable multiplicity and fragmentation of society by the encouragement and provision of a system of person-centered, creative, and democratic education, which is expected both to extend and strengthen the sense of solidarity in Swedish society.

Most narratives about the future being created in the present are created from an early twenty-first-century perspective with focus on the complexity of today's problems. In different contexts and with different logics, we are told the same story, that is, that we have to live with constantly rapid changes in society and with conflicts between various group interests and forces in society. The Government Commission, *Health on Equal Terms—National Targets for Health*,[27] which is proposing national targets for public health in Sweden, is concerned about the effects of changes in society and the present health situation in Sweden and about what will happen in the future, if appropriate measures are not taken. According to the Commission, the Swedish model—a welfare state and with a public health system—is "exposed to huge external and internal tensions."[28] Increasing inequalities with regard to health and social conditions threaten the basic trust in society and the attempt to found a society based on solidarity between different population groups.

A weakening of fellow-feeling and an increasing lack of trust lessen the possibilities of finding solutions based on a sense solidarity that provides welfare to everybody.[29]

In a deeper meaning the Public Health narrative is not about health. First and foremost the text, and thus the narrative, is about Society, Man, and Future. A similar picture of the future is painted by the

Teacher Education Commission,[30] commissioned to draw up the design required for the teacher of tomorrow. According to the Commission the rapid transformation of society makes it impossible to predict what knowledge and skills students will need in the future.

For that reason, an essential dimension of teacher competence is an ability to foresee and manage changes and to design a learning environment that enables development of basic human relationships.[31]

A future-oriented teacher is thus produced, a teacher with the competence to work in a pluralistic and constantly changing society, and equipped with an ability to support children in their learning of how to understand and manage life and society of the future. According to this narrative the future is already here and there is nothing to wait for. You cannot really be certain what the future holds, but this is of no interest since the future is shaped by what we do and how we are preparing and mobilizing ourselves in the present through "the choices we are making *today*, which are decisive for the development of an improving future. We can not sit and wait."[32] According to the Democracy Report, in this constantly changing world, it is "important as a matter of urgency to reinforce the kind of democratic disposition that can create democratic institutions with the power of solving problems in a peaceful way."[33] What should be defined as common goals and solidarity, however, is one of the problems for the Committee Report and other similar works on reformation at the present. The notion of pluralism and the contemporary trend based on a culture of networking are somewhat problematic with regard to this. The emergence of pluralism is, on the one hand, favored because of its spontaneous nonhierarchal way of functioning and for the way it facilitates active participation and spontaneous initiatives. On the other hand, it runs the risk of being developed in an unintended direction and consequently developing into something that becomes ungovernable since there is no longer a system of comprehensive or coherent schooling in the principles and practice of democracy and solidarity as was the case in earlier periods.

The expansion of the culture of networking challenges the democracy of popular movements in several respects. It shapes the conditions into becoming more conducive to rapid initiatives and to more action-oriented participation, which is favorable for those taking part. On the other hand the popular movements are a training ground for the exercise of democracy for its members.[34]

The Committee Report sees this trend of civil networking and net-widening and the increasing cultural multiplicity and fragmentation, as a problem of the future, notwithstanding that at the same time this trend is perceived to be the very source of the possibility of widening the notion of solidarity and an increasing sense of freedom. Dialogue seems to be the saving dimension of the current mode of democratic problem-solving.

Recognize, accept, support, and stimulate the free associations of citizens! Invite them to recurrent conversations about the long-term creation of the cement that deepens the democracy.[35]

The Committee Report is a rather characteristic manifestation of the kinds of reasoning taking place in the circles of government of the early twenty-first century. The motto is: Since the democratic society is developing in rather indefinite and uncertain ways we have to speak to each other. The imagination of an uncertain future requires the citizens to enter the path of communication and meeting others in interaction, to entertain the possibility of creating spaces of autonomy and self-regulation in the realm of an extended sense of community. And Dewey's figures of thought are thus inscribed in technologies created to resolve the paradox of a widening of the notion of solidarity on the one hand, and a continuing movement in the direction of an increasing sense of freedom, on the other.

Governance through Interaction and Communication

A prominent technology for the development of autonomy and self-regulation is the creation of spaces of interaction and communication to widen the sense of community by including and encouraging the integration of diverse values and lifestyles. Dewey's ideas about democracy, community, and communication, thus, are directly and indirectly inscribed in a number of areas of political discourse. We demonstrate this with regard to the areas of Education, Public Health, and Criminal Justice in the following sections.

Education

The importance of the assignment to develop democratic competence in children and young people in education has been emphasized in the debate about education during the past decade.[36] One crucial aspect of this technology is the ability to engage in deliberative democracy and deliberative discussion. In this context, Dewey's figures of thought have been inscribed on the political discourse on education both in the name of Dewey himself and in the name of other thinkers such as Benhabib, Habermas, Mead, and Putnam. According to the current narrative, school should be a meeting place where children, young people, and adults from a variety of cultural backgrounds and experiences can gather and, through communicating, have their minds opened to the enriching experience of exposure to and interaction with others in a context of diversity.

Communication requires of those participating to possess an ability to listen, to consider their own view and that of others. It requires people to make a case for their own standpoint whilst at the same time working with others to find common ground on norms and values.[37]

The main goal of this kind of deliberative communication, however, is not about the individual; the goal is about development of school as a democratic community and, in the final analysis, about the development of society as a whole.[38] In this narrative democracy becomes a moral ideal to be practiced in everyday life. An ideal that is similar to Dewey's thought on democracy.

> . . . democracy is a personal way of individual life; (that) it signifies the possession and continual use of certain attitudes, forming personal character and determining desire and purpose in all the relations of life.[39]

Thus, Dewey's way of thinking about democracy is embedded in contemporary practices of governance and in the contemporary project of developing a democratic and communicating citizen taking responsibility for his/her own life and that of others, as well as for the democratic development of society as a whole. The collective endeavor to find common ground with regard to values and norms becomes a technology for the management of potential social conflict and to create feelings of solidarity in a pluralistic and constantly changing society.

Like Dewey,[40] Skolverket[41] is arguing for a non-dualistic relation between knowledge, values, and social processes; these develop in the interaction between human beings in different social contexts. Nevertheless, there are clear limitations to what can be questioned within the practice of communication of the kind we are discussing here. Deliberative communication "does not mean that all opinions and perspectives can be tolerated."[42] You cannot make compromises regarding the basic values settled in the curriculum.

The basic values decided on have to be valid. The basic values in the curriculum must not become relative.[43]

There are thus limitations on the inscription of Dewey's nondualistic way of thinking with regard to the discourse on education. According to Skolverket, there are absolute and noncontingent values and norms that cannot be questioned, such as freedom, solidarity, and equality. In the final analysis, deliberative communication and democracy become the technology for passing on specific values and capabilities to children and young people. The focus on deliberative communication also has an impact on the education and training of teachers. The aspiring teachers must be able to design the educational setting as a social meeting place and a learning organization, as a technology aiding the promotion of deliberative communication and democratic competence. Hence, the aspiring teacher has to have democratic competence and share the basic values inscribed in the curriculum.

The teacher has to be aware of the basic values that must be taught in school and have the ability to convey and how to instil these values in the minds of children and young people.[44]

As we already have seen, these basic values are considered as absolute and noncontingent and as a foundation of society. These values cannot be questioned; the teacher's task is to convey the values to the pupils and make them shared values, through dialogue, discussions, and reflection, because this method is regarded as being the most likely to succeed. Thus, the teacher of tomorrow, in the process of preparation, is a reflecting, communicating, and analytic practitioner with an ability to design learning environments. To this end, the teacher training context needs to function as a deliberative democracy focusing on deliberative communication. The student teacher has to be a coactor with a genuine sense of shared responsibility for most activities in the context of teacher education and training.

However, this is not only a student privilege; in a broad sense, the student is obliged to participate in the internal activities of the university. Higher education means, not only studying for exams, but also to be a coactor and coresponsible for the universities' internal life as well.[45]

In the conception of the Commission, engaging in deliberative communication and exercising freedom become obligations for aspiring teachers. This thinking is based on the assumption that the individual is positioned as the cocreator of his/her future and that the future must be freed from earlier meritocratic thought and allowing the possibility of cocreation on an equal basis, that is, it presupposes a new and more flexibly organized future built around a radical democratic power.

Public Health

The Public Health Commission's vision for the future is a society with robust social cement holding it together, a society where solidarity is central, where trust in society and in other people prevail.[46] The key element in making this vision come true is a commitment to develop and deepen democracy and participation in society as a whole. In the staging of this project of democracy a community approach becomes crucial.

Health is created and lived by people within the settings of their everyday life; where they learn, work, play, and love.[47]

The key principle for government in this approach is to mobilize local multi-organizational alliances labeled as, for instance, health-promoting communities, health-promoting workplaces, health-promoting schools. The aims are to strengthen solidarity and social relationships through communication, to develop new knowledge about health issues, and to increase the sense of commitment among all interest groups to share in a common goal, which is to develop the local community as a health promoting society. Thus, in contemporary Public Health discourse there is a deep, more or less Deweyian, belief in face-to-face cooperation and communication as a way of managing conflicts and disputes between group interests in society.[48] Applied to school as a community for health promotion, the idea is to involve teachers, students, and other members of the staff, parents, citizens, voluntary organizations, private bodies, community groups, and other local agencies in the work for school as a health promoting organization. In Public Health discourse school becomes a community, and the concepts of school and society become interchangeable in the context of Public Health discourse. Through the creation and regulation of new relationships in a space that is local Dewey's figures of thought form an element in a technology for governing in the name of public health and for the development and fabrication of the health-promoting citizen, who is prepared to self-govern, taking responsibility for him/herself and his/her environment.

The individual is seen as an active, acting, and responsible social being who strives to shape his conditions of life in accordance with his desires and needs, but who also strives to take responsibility for society as a whole and the well-being of others.[49]

In this context, Dewey's figures of thought are inscribed in a traveling library including ideas from thinkers such as Antonovsky, Putnam, Rawls, and Sen.

The paradox of the forces of individualization on the one hand, and the need of cooperation on the other, is solved by the Commission while introducing the concept of *the solidarity-feeling individualist*, who as an active, free, and autonomous individual takes responsibility for himself and for others and for creating an environment supportive of public health and the development of communication and democracy in society as a whole. The crucial welfare dimension and the ultimate criteria for success in the public health arena are, according to the commission, "human beings' freedom to control and arrange their conditions of life in accordance with their conscience" and in cooperation with others.[50] Thus, the freedom of individuals requires cooperative actions and participation in projects intended to make life and health better for all. Dewey's thought is present in this concept of freedom, for according to Dewey, freedom requires pluralism, which can only be guaranteed by cooperative actions. Thus, freedom is included in the concept of community.[51]

Criminal Justice

Dewey's ideas about community and communication are also present in changing contemporary ideas about "doing justice." In this field the local community approach operates in the name of *restorative justice* or *community mediation*.[52]

> . . . mediation provides a practical conceptualization of how the self will have to relate to itself to appropriate a style of life, a mode of being, which will avoid future disputes. It is this dimension of the telos that leads advocates to view mediation as an educational process, as a means of imparting a "life skill" to selves, and as a way of avoiding future conflicts.[53]

The principles of restorative justice emphasize the importance of problem solving and communication and interaction between parties involved in and affected by crime or other conflicts, and they focus on what is to be done in the future, not on what has been done in the past.

> Restorative justice places both victim and offender in active problem-solving roles that focus upon the restoration of material and psychological losses to individuals and the community following the damage that results from criminal behavior. Whenever possible, dialogue and negotiation serve as central elements of restorative justice. [. . .]

Problem solving for the future is seen as more important than establishing blame for past behavior. Public safety is a primary concern, yet severe punishment of the offender is less important than providing opportunities to: empower victims in their search for closure and healing; impress upon the offender the human impact of their behavior; and promote restitution of the victim.[54]

Restorative justice, and community approaches within education and public health, could be seen as new ways of ordering the lives of citizens in contemporary societies. This is based on the rationale for promoting communication, that is, we make peace by speaking to each other and we become reconciled with each other by telling the "truth" about ourselves. This new way of doing justice is very much about interaction, whilst at the same time it serves an ambition to personalize and humanize the justice process in order to facilitate "the empowerment of both parties to resolve the conflict at a community level."[55] In the same way as the institutional practices of education and public health pay tribute to concepts such as empowerment, future orientation, interaction, communication, democratization, and so on, similar Deweyian figures of thought are operating in the realm of restorative justice. On one level its official aim is to resolve conflicts between people and increase harmony, on another level it is a question of producing new technologies for the governing of *the soul*.[56]

The Swedish experiment of community mediation started in 1978. It was based on the collaboration between the police, the legal system, and social services, and has during the early 2000s expanded into a rather widespread activity. On the one hand it is an extension of the legal system, and on the other, it is also a new way of linking the citizen closer with the community.

The theory of restorative justice also contributes to the will of empowering the local community and strengthening the local influence over the individual by moving the legal system to a lower level—one is of the opinion that misconduct and wrong behavior should be corrected in a less formal way, namely through social control at a local level.[57]

In this context "the state" is no longer imagined as the victim; "restorative justice theory postulates that criminal behavior is first and foremost a conflict between individuals."[58] One of the aims of mediation is to get the offender to express remorse and ask for the victim's forgiveness. The victim is invited to listen, and to understand the whole story of the offender's criminal activity, and the story has to be told in the presence of the victim and the mediator/confessor. In this respect, the mediation process is the technology for the empowerment of individuals in conflict through a confessional face-to-face relation.

The Learning Society

The epistemological basis for government action is that it is possible to order and control the future in the present by qualifying and preparing the individual citizen with dispositions for new commitments in a constantly changing, pluralistic, and fragmented world. To make this way of governance possible, lifelong and life-wide learning becomes crucial.[59] The citizen promoted in all narratives about the future has to be an educational and learning subject. A prominent Scandinavian educationalist is emphasizing this development in visionary words:

> We are entering the era of the knowledge based society where the speed of change is ever greater and where demands for new skills and qualifications are made on each of us. Education will no longer be something which is linked to a certain age or educational background, but will be a necessity and a self-evident part of everyday life for citizens from all walks of life, social classes and occupational groups. We are already there. We know that "life-long learning" has become a reality.[60]

Learning is not limited to the setting of a classroom or to a specific place or time. Education and training need to extend to encompass the entire body of society and to be an ongoing, everlasting pursuit.

It is not possible for the education system to carry out this task alone.[61]

The subject citizen must be prepared to learn during all of his/her life and to engage with learning in a wider sense. Education is once again a project for national mobilization, but with a hugely different meaning than earlier. Education is not just a question of schooling, but something that permeates the governance of all social activities. The new technologies applied in the area of criminal justice have more pedagogical implications than penal ones. It is more about learning how to be a law-abiding, problem-solving, communicative, and responsible member of society than it is to do with punishment for wrongdoing.[62] In the Public Health field technologies of pedagogy become significant for the conduct of the health of the population and the conduct of the individuals, not only within formal educational settings, but also in people's everyday lives and in society as a whole. The Public Health

Committee emphasizes that it is "important for a society that the citizens look upon learning and personal development as a life-long process."[63] According to the Commission on Teacher Training and Education[64] learning is an integral part of all activities in society. Society is seen as a learning society. The Commission emphasizes that school and teacher education as social institutions in many respects have a potential to become models for social life in general.

There is a paradox involved here. On the one hand, it rather seems to be the case that educational thought is spreading and tending to dominate even more spaces than in the past. On the other hand, educational thought is currently tending toward becoming narrower than before making pedagogy limited and reserved for the realm of the individual as an educable subject and to an individual task, often expressed by the rhetoric of lifelong and life-wide learning, self-regulation, empowerment, and so forth, indicates a more complex view. It does not seem farfetched to argue that society has turned into a school. Thus, Dewey's notion of "School as Society" has been reshaped into "Society as School."

Conclusion

Dewey's figures of thought have been inscribed into the Swedish political context since the end of the nineteenth century. His ideas of democracy, community, and communication have been used, directly in the name of Dewey and indirectly in the names of others, when changing conditions and concerns about the future have put the Future, Society, Citizen, School, and the Child on the stage. However, it is not the same conception of Dewey that has been brought into play in different political narratives, and it is of course not the same Future, Society, or Citizen either. From a genealogical point of view it is, furthermore, relevant to say that the indigenous foreigner Dewey even was here before he arrived, as a figure of thought and discursive practice. One example is Berg, who in the beginning of the twentieth Century was inscribing Dewey into his narrative about a school promoting the development of citizen capable of fellow-feeling but not questioning existing power relations in society. During the 1930s and the 1940s, Dewey was operating in socialist narratives and visions to alter existing power relations in society. Now, Dewey's thought became embedded not only in the development of the postwar Swedish school system but also in the formation of the Swedish welfare state, the People's Home, as a whole. In this project school was seen as a crucial institution for practice in the constitution of a future-oriented and democratically minded citizen. The vision of the People's Home was built on the idea that it was possible to foresee and make plans about the future, and that it was possible to fabricate autonomous individuals, with a strong sense of individuality, as a part of a collective nation-building enterprise.

Compared to the nineteenth century and the first half of twentieth century, there has occurred a shift in the thinking and practice within such institutional domains as public health, teacher education, and criminal justice, and with regard to organizing the division of responsibilities and duties between citizen and state. In this respect, we see today signs of a new recodification of the concept of freedom. As Wagner[65] claims, this is a significant event, which can be seen as an expression of the changing view of the welfare state and of how the citizen should be governed. Wagner argues that the welfare states are no longer able to structure their citizens' lives in the name of their collective welfare, and that welfare tasks, to an increasingly greater extent, have become the duty of the individual citizen. The idea of the citizen as cocreator of the future is not a new one, but Wagner's point is that the liberal way of thinking has radicalized recently. The future is controlled less through society and more through individuals' own initiatives. A change of language corresponds to this liberalization in governing as does a change in the categorizations, and divisions of institutions and areas of activity.

Today, Dewey is newly arrived in this new constructivist narrative about Society, the Future, and the Citizen. In a constantly changing society it will not be possible to foresee and plan for the future. The individual subject, according to this narrative, are not as stable and predictable as before, but capable of problem solving, of taking responsibility, and being flexible, and so being capable of responding to changing conditions and situations. To make this reconstruction possible, lifelong as well as life-wide learning becomes crucial.

Paradoxically, the citizen is governed with the object of becoming a learning individual, at the same time as educational thought and pedagogy is applied over the entire body of society. School becomes society and society becomes school.

Yesterday's narratives using Dewey's figures of thought are quite different from contemporary ones when it comes to the question of organizing the relationships between Man, Society, and Future. However, they are all variations on the same theme, that is, a variation of the problematic question of "who the citizen and the child are and should be" (Popkewitz, chapter 1, in this book) and the question of the will to govern in the name of the Future.

Notes

1. Thomas S. Popkewitz, "Dewey and Vygotsky: Ideas in historical spaces," in *Cultural History and Education: Critical Essays on Knowledge and Schooling*, ed. Thomas S. Popkewitz, Barry M. Franklin, and Miguel A. Pereyra (New York and London: Routledge Falmer, 2001), 313.
2. Per Albin Hansson, *Folkhemstalet* (Stockholm: Sweden: Riksdagstrycket, 1928).
3. Oskar Olsson, *Demokratins skolor* (Stockholm, Sweden: Frihetens Förlag, 1943).
4. John Dewey, "Democratic ends need democratic methods for their realization," in *The later works (1925–1953)*, ed. Jo Ann Boydston, Vol. 14, s. 224–230 (Carbondale and Edwardsville: Southern Illinois University Press, 1939b), 367.
5. Ernst Wigforss, *Skrifter i Urval. VII; VIII. Minnen* (Stockholm, Sweden: Tidens Förlag, 1980).
6. Ernst Wigforss, *Skrifter i Urval. VII; VIII. Minnen*, 327.
7. Alva Myrdal and Gunnar Myrdal, *Kontakt med Amerika* (Stockholm, Sweden: Bonnier, 1941), 90.
8. Myrdal and Myrdal, *Kontakt med Amerika*, 96.
9. SOU 1948: 27, års skolkommissions betänkande med riktlinjer för det svenska skolväsendets utveckling (Stockholm. Sweden: Eckleastikdepartementet, 1946).
10. Fridtjuv Berg, *Folkskolan såsom bottenskola: Ett inlägg i en viktig samhällsfråga* (Stockholm, Sweden: Lars Hökerbergs Förlag, 1883).
11. Berg, *Folkskolan såsom bottenskol*, 60–61.
12. SOU 1948: 27, *års skolkommissions*, 5.
13. SOU 1948: 27, *års skolkommissions*, 4.
14. John Dewey, *Democracy and Education* (New York: The Free Press, 1966/1918), 98.
15. Sven G. Hartman, Ulf. P. Lundgren, and Ros Mari Hartman, *Individ, skola och samhälle. Pedagogiska texter av John Dewe* (Stockholm, Sweden: Natur och Kultur, 2004).
16. Yvonne Hirdman, *Att lägga livet till rätta: Studier i svensk folkhemspolitik* (Stockholm, Sweden: Carlssons Förlag, 1989).
17. See Ulf P. Lundgren, *Att organisera omvärlden: En introduktion till läroplansteori* (Stockholm, Sweden: Liber, 1988).
18. See Kenneth Petersson, *Fängelset och den liberala fantasin: En studie om rekonstruktionen av det moraliska subjektet inom svensk kriminalvård* (Norrköping, Sweden: Kriminalvårdsstyrelsens forskningskommitté, 2003).
19. See Ulf Olsson, *Folkhälsa som pedagogiskt projekt: Bilden av hälsoupplysning i statens offentliga utredningar* (Uppsala, Sweden: Studies in Education, 1997).

20. Kenneth Hultqvist, "The travelling state, the nation and the subject of education," in *Dangerous Coagulation: The Uses of Foucault in the Study of Education*, ed. Bernadette M. Baker and Katharina E. Heyning (New York: Peter Lang, 2004).

21. Kenneth Petersson, Ulf Olsson, Thomas S. Popkewitz, and Kenneth Hultqvist, *Reframing Educational Thought: Subjects and Technologies of the Future in the Early 2000* (Paper for ECER Conference, Crete, September 22–25, 2004).

22. The title is borrowed from Lena F. Ljunghill, "The future is here and now—what are we waiting for!" *Pedagogiska magasinet* 1, no. 1 (Stockholm, Sweden: Lärarförbundet, 1996).

23. Lena F. Ljunghill, "The future is here and now—what are we waiting for!" 7.

24. Skolverket, *Det livslånga och livsvida lärandet* (Stockholm, Sweden: The National Agency for Education, 2000a).

25. SOU 2000: 1, *En uthållig demokrati: politik för folkstyrelse på 2000-talet, Demokratiutredningens betänkande* (Stockholm, Sweden: Fritzes Förlag).

26. SOU 2000: 1, *En uthållig demokrati*.

27. SOU 2000: 91, *Health on Equivalent Conditions: National Targets for Health* (Stockholm, Sweden: Socialdepartementet).

28. SOU 2000: 91, 55.

29. SOU 2000: 91, *Health on Equivalent Conditions*, 82.

30. SOU 1999: 63, *Att lära och leda: En lärarutbildning för samverkan och utveckling, Lärarutbildningskommitténs slutbetänkande* (Stockholm, Sweden: Fritzes Förlag).

31. SOU 1999: 63, *Att lära och leda*, 57.

32. Per Dalin, *Utbildning för ett nytt århundrade*, Bok 1 (Stockholm, Sweden: Liber, 1994).

33. SOU 2000: 1, *En uthållig demokrati: politik för folkstyrelse på 2000-talet, Demokratiutredningens betänkande* (Stockholm, Sweden: Fritzes förlag), 204.

34. SOU 2000: 1, *En uthållig demokrati*.

35. SOU 2000: 1, *En uthållig demokrati*, 206–208.

36. Skolverket, *En fördjupad studie om värdegrunden: om möten, relationer och samtal som förutsättningar for arbetet med de grundläggande värdena*. Dnr 2000: 1613 (Stockholm: Skolverket, 2000).

37. Skolverket, *En fördjupad studie om värdegrunden*, 6.

38. Tomas Englund, *Deliberativa samtal som värdegrund: historiska perspektiv och aktuella förutsättningar* (Stockholm, Sweden: Skolverket, 2000).

39. John Dewey, "Creative democracy the task before us," in *The Later Works (1925–1953)*, Vol. 14, ed. J. A. Boydston (Carbondale and Edwardsville: Southern Illinois University Press, 1939), 224–230.

40. Dewey, *Democracy and Education*, 98.

41. Skolverket, *En fördjupad studie om värdegrunden*.

42. Skolverket, *En fördjupad studie om värdegrunden*, 10.

43. Skolverket, *En fördjupad studie om värdegrunden*, 10.

44. SOU 1999: 63, *Att lära och leda*, 10.

45. SOU 2000: 91, *Health on Equivalent Conditions*, 63, 297.

46. SOU 2000: 91, *Health on Equivalent Conditions*.

47. "Ottawa Carter WHO/1987," in *The Health Promoting School: Policy, Research and Practice*, ed. S. Denman, A. Moon, C. Parsons, and D. Stears (London: Routledge, 2002).

48. cf. John Dewey, "Creative democracy the task before us."

49. SOU 1999: 137, Mot nationella mål för folkhälsa (Stockholm: Socialdepartementet), 6.

50. SOU 2000: 91, *Health on Equivalent Conditions*, 63.

51. Dewey, *Democracy and Education*.

52. Mark S. Umbreit, *Victim Meets Offender: The Impact of Restorative Justice and Mediatio* (New York: Willow Tree Press, 1994); Howard. Zehr, *Changing Lenses. A New Focus of Crime and Justice* (Scottsdale: Herald Press, 1990); Mark S. Umbreit, *The Handbook of Victim Offender Mediation: An Essential Guide to Practice and Research* (San Francisco, CA: Jossey-Bass, 2001); R. Bush, A. Baruch, and J. P. Folger, *The Promise of Mediation: Responding to Conflict through Empowerment and Recognition* (San Francisco, CA: Jossey-Bass, 1994).

53. G. Pavlich, "The power of community mediation: Government and formation of self-identity," *Law & Society Review* 50, no. 2 (1996): 725.

54. Umbreit, *Victim Meets Offender*, 2.

55. Umbreit, *Victim Meets Offender*, 17.

56. Nikolas Rose, *Governing the Soul: The Shaping of the Private Self* (London, New York: Free Association Books, 1999).

57. Brottsförebyggande Rådet (The National Council for Crime Prevention), *Medling vid brott: Slutrapport från en försöksverksamhet*. Brå-rapport 1999: 12 (Stockholm, Sweden: Fritzes, 1999) 11.

58. Umbreit, *Victim Meets Offender*, 2.

59. SOU 2000: 1, *En uthållig demokrati*.

60. Per Dalin, *Utbildning för ett nytt århundrade*, 143.

61. Per Dalin, *Utbildning för ett nytt århundrade*, 101.

62. Umbreit, *Victim Meets Offender*.

63. SOU 2000: 91, *Health on Equivalent Conditions*, 423.

64. SOU 1999: 63, *Att lära och leda*.

65. Peter Wagner, *A Sociology of Modernity: Liberty and Discipline* (London, New York: Routledge, 1994).

3

Langue as Homeland: The Genevan Reception of Pragmatism

Daniel Tröhler

Hardly in any other European country than in Switzerland can we demonstrate evidence of an early and long-lasting interest in pragmatism.[1] To be accurate, the Swiss reception had three centers, one of them a kind of "headquarters": Neuchâtel, Lausanne, and above all Geneva—in other words, in the three Protestant capitals of cantons in the French part of Switzerland. In order to explain this phenomenon, I first reconstruct the early contacts between Geneva and the United States; these show a familiarity going beyond personal sympathy. Next, I focus on the transcontinental exchange after 1905, which became more and more educational, in connection with progressive education; this points out the core element of the Genevans' specific interest in Deweyan education: activity. Then I try to show that this core element was part of the liberal reformist *langue*, showing that *langues* are homelands without political borders, although they do not rule out local readings and accentuations. Finally, I show that the activity of reception was not limited to the Genevans, and that they, in turn, experienced an active reception abroad, showing that libraries do not travel without specific preconditions in the destinations.

"C'est mon homme"

The transcontinental exchange of ideas between Geneva and the United States that proves to be important for our topic began—as far as reconstruction is possible—with the publishing of contributions by

William James in the dominant British journal *Mind* and in the French journal *La Critique philosophique* in the 1880s. Apparently, the Genevan scholar Théodore Flournoy, who was twelve years younger than James, showed a deep interest in James's research on physiological psychology, including James's response to Huxley's famous 1874 statement "We are conscious automata"[2] and James's reflections on emotions.[3]

On one of James's long trips to Europe, he visited the International Congress of Physiological Psychology in Paris in 1889, where he was introduced to Théodore Flournoy.[4] A year later, James sent his Genevan colleague his two-volume work, *Principles in Psychology* (1890),[5] and Flournoy expressed his deep appreciation in a letter dated October 15, 1890, thanking James for the books and confessing that, besides a few minor points of disagreement, James inspired him and put into words his own feelings: ". . . I am forced frequently to say, in speaking of you, 'C'est mon homme!.' "[6] In the spring of 1891, Flournoy published a most favorable review of *Principles* in the *Journal de Genève*, and James assured him in a letter on March 31, 1891 that Flournoy "had better than anyone else caught the 'point of view' of my lengthy pages." Reading Flournoy's *Métaphysique et psychologie* (1890)[7] James felt the transcontinental affinity even more deeply. James expressed his hopes that, with each other's help, "our 'School' will prevail!"[8]

This "school" was not based only on personal characteristics or intellectual affinities. James and Flournoy both felt "at home,"[9] because they shared the same mode of thinking, the same intellectual horizon that was shaped by the Calvinist-Protestant and republican *langue*.[10] It is not by accident that James calls Switzerland, foremost Geneva, "the terrestrial paradise" and praises Flournoy as "a citizen of that fortunate republic." According to James, the worldwide "neurotic *fin-de-siècle* element" had "comparatively little hold on" Geneva. This was a compliment to Geneva, but also an inexplicit side attack on Germany, with its constant prejudices against the United States (31). Nowhere (at least not in Europe), says James, have the beauty of nature and the political institutions joined together more harmoniously than in Switzerland (115). Discussing the intervention of American troops in Spanish-occupied Cuba, Flournoy wrote to James on December 11, 1898 in the same "spirit" but more explicitly: ". . . on the whole we [the Swiss, DT] applaud the triumph of a republican and Protestant nation representing liberty and modern civilization over the old monarchical and Catholic despotism" (76).

But neither political nor religious topics dominated the discussions between the two scholars. Their correspondence focused on personal subjects and, in second place, on the evolving world of academic psychology. Flournoy and James had their own laboratories in their universities, which very soon began to trouble them both. The enthusiasm that Wilhelm Wundt's physiological psychology had been able to engender in the 1870s was eclipsed by the slow and uninteresting results. Flournoy talks about "My laboratory, which bores me more and more" and says that the results "accomplish nothing worthwhile" (December 30,1892, 17);[11] James confirmed that Wundt's time was over and that he should be retired (20). In 1895, Flournoy began to suffer even, because his laboratory was "becoming a fixed, morbid idea, a real phobia with me" (45). In 1896, James's judgment was clear: "The results that come from all this laboratory work seem to me to grow more and more disappointing and trivial" (61). These lines were written in the same year that John Dewey's *The Reflex Arc Concept in Psychology*[12]—based on the laboratory work at the University of Chicago under the direction of James Roland Angell—brought to public expression the unsatisfactory results of a mechanistic and dualistic psychology.

Psychology, Liberal Reformist Protestantism, and Evolution

The skepticism against the laboratory work was rooted in religious convictions—but *not* in theological questions. The distance that William James felt can be demonstrated as early as 1879 in his challenge of Huxley's mechanistic Darwinism, which Flournoy applauded in Geneva long before he and James met in Paris for the first time. Among many other arguments, James cites a strophe of a poem by Goethe (in German):

> Nur allein der *Mensch*
> Vermag das Unmögliche.
> Er unterscheidet, wählet und richtet,
> Er kann dem Augenblick
> Dauer verleihen.[13]

The poem, *Das Göttliche* (The Divine), was published in 1783 and was one of the most popular poems of the late eighteenth century. At first glance, that James would cite a poem about the divinity of humankind is surprising in the context of a scholar considered to be an evolutionist. But it is a common error to interpret the intellectual discussions in the late nineteenth century as similar to the intransigent debates between creationists and evolutionist today. Finding phenomena like the struggle for existence, natural selection, probabilities, or adaptation convincing did not necessarily mean that you had no faith or were unchristian. "True" science was not meant to be opposed to Christianity, as the history of the universities in the United States shows.[14] Menand says about James: "He was Darwinian, but he was not a Darwinist. This made him truer to Darwin than most nineteenth-century evolutionists."[15]

James took a position that was shared by many American intellectuals in the last third of the nineteenth century. It was called *liberal*, and liberal meant not caring much about dogmas, such as the idea of original sin. Theological arguments could not, in their view, add anything to the only important thing, namely, *being* religious, *experiencing* religion, in a Protestant sense. Of course, the theological tensions between Darwin(ism) and the belief in creation were recognized, but there were efforts to reconcile them, *pragmatically*. This pragmatic handling of questions of dogma was deeply rooted in the liberal Calvinism of the end of the nineteenth century: Religion is a (social) fact, and theoretical discussions about it are rather idle.

"True" life was religious as well as social and political—a deep Calvinistic conviction—and was therefore not to be understood by mechanistic methods. That is why both James and Flournoy were interested in doing psychical research in spiritism, telepathy, and hypnosis, and in working with mediums. Flournoy's most famous book, *Des Indes à la planète Mars: étude sur un cas de somnambulisme avec glossolalie* (From India to the Planet Mars: A Study of a Case of somnambulism: with glossolalia), was published in 1899 and translated into English as early as 1900, published in Italian in 1905, and in German in 1914. There were countless new editions and reprints of the book throughout the twentieth century—the last printing in English was in 2003 and in French in 1994. Years before this publication, William James was attending (anonymously) the séances of medium Mrs. Leonora E. Piper, and he wrote about her in *The will to believe*.[16] Apparently, James had an agreement with his Cambridge colleague and psychical researcher Frederic W. H. Myers; they agreed that when the first of them was dying, he would send news to the other from the kingdom of the unknown. And indeed, in 1901 James was waiting, pen and notebook in hand, outside the door of the hotel room in Rome where Myers lay dying. James' notebook remained empty, as Alex Munthe tells us in his immensely popular *Story of San Michele*.[17] Spiritism remained an important topic in the correspondence between James and Flournoy up to James's death in 1910. Visible life was understood to be as real as religious feelings, and reflection about psychological facts was called "philosophy." By 1900, religion had become James's central topic. His *Varieties in religious experience*[18] was translated into French in Switzerland by a friend of Flournoy's, Frank Abauzit, and edited in Paris with an introduction by French philosopher Emile Boutroux as *L'expérience religieuse: essai de psychologie descriptive*.[19] It is no coincidence that the Faculty of Theology of the University of Geneva had planned to award James the honorary degree of "Docteur en Théologie honoris causa" in 1909[20] or that the 1910 lecture that Flournoy was to hold on William James was organized by a union of young (Protestant) Christians in Neuchâtel. Religion was questioned as metaphysical *humbug*, but not as a real phenomenon.

Flournoy wanted to understand religion psychologically, for example, through scientific methods, starting out from the fact of the diversities of religious experience in different social contexts.[21] Religious psychology, he says, has the advantage over religious philosophy, as it restrains us from giving any interpretation of what is true—and he remembers history "where every one believed himself to be the only judge not only for himself, which would be perfectly legitimate, but for the others, too."[22] This speaks quite strictly against Catholicism, against Lutheranism—but not against Protestantism as it is found in Baptist circles, for instance. What seems to be secular proves to be—on a quite invisible background—liberal reformist Protestantism. Flournoy published this 1903 article, *Les principes de la psychologie religieuse* (The Principles of Religious Psychology), in the journal *Archives de psychologie*, which he and his nineteen-year-younger cousin, Edouard Claparède, had coedited since 1901. Claparède eventually became a professor of psychology at the University of Geneva (1908) and founded the "International Bureau of Education"[23] in 1925 together with Pierre Bovet. Bovet was the author of *Le sentiment religieux et la psychologie de l'enfant* (The Religious Feeling and Child-Psychology),[24] third of the prominent Genevans in the reception of pragmatism, Adolphe Ferrière, took the chair of the International Bureau of Education in 1929 and was later succeeded by his "disciple" Jean Piaget, whose first writings had also dealt with religious feelings. But by then, Pragmatism and Dewey had become an essential element of the intellectual discourse in Geneva. The questions to be examined are how Pragmatism was understood and, vice versa, what reading of Pragmatism was able to express what the Genevans were thinking.

Pragmatism and Progressive Education

The issue of the transcontinental discussions was "facts *and* religion," as William James pointed out in his public lectures in Boston and New York 1906 and 1907, published under the title *Pragmatism. A New Name for Some Old Ways of Thinking; Popular Lectures of Philosophy*.[25] He thus challenged both of the dominant doctrines of his time, rationalism and empiricism. "My reading is more philosophical than psychological in these days," James writes on January 2, 1907 to Flournoy.[26] It is as if the Genevans were just waiting for this philosophical justification of their interest in the American psychology. Flournoy confirmed that he saw James as the "genuine creator" of pragmatism, "for the worthy Peirce does not appear to me to have been more than the starting push" and that without James, Dewey, and Schiller, neither Bergson nor Boutroux would have been able to develop their philosophical systems.[27] Dewey may have been received in Geneva earlier, but it is certain that he was introduced in a letter of January 1, 1904, when James highly recommended Dewey's *Studies in Logical Theory*[28] and sent along a copy.[29] In a letter of 1907, Flournoy writes to James: "I have grown more and more deeply into pragmatism, and I rejoice immensely to hear you say: 'je m'y sens tout gagné.' It is absolutely the only philosophy with *no* humbug in it, and I am certain that it is *your* philosophy."[30] In the field of psychology, pragmatism seemed to be the tool against causal empiricism, and in the field of philosophy it was the tool against (German) rationalism or idealism. Together with Bergson, James believed, his pragmatism would "converge [history] toward an antirationalistic crystallization. Qui vivra verra!"[31] the context of this mission, James praises Flournoy's cousin Edouard Claparède—and at the same time makes an accusation against German philosophy: "When will the Germans learn that art?"[32]

With James's (re)turn to pragmatism in the early 1900s, the intellectual dispositions between Geneva and Harvard had been adjusted and clarified to the extent that now, James, Dewey, and his (former) Chicago colleagues could be received well beyond philosophical reflections on religious psychology. Pragmatism struck the right note for the mental dispositions of an old republic, which had demanded political ethics, virtue, and education ever since Rousseau's *Lettre à d'Alembert* (1758). James's *Talks to teachers* appeared in French translation in Lausanne under the title *Causeries pédagogiques* in 1907,[33] which launched a broad discussion within the field of education. After having announced the French translation already in February 1907, in March the Department for Public Education in Neuchâtel lauded the advantages of James's psychology for education, not forgetting to mention the difference between psychology and education,[34] in the *Bulletin Mensuel*, a newsletter that was distributed to all school board members, school administrators, and teachers. A month later, the Lausanne journal *L'Educateur* gave a brief summary of the translation,[35] and in August, the same journal contained a short introduction to the philosophy of William James written by his translator, Louis[-Samuel] Pidoux.[36] In 1910, the educational society of Neuchâtel dedicated its fiftieth anniversary celebrations to James. Pierre Bovet, since 1903 professor of philosophy in Neuchâtel, dedicated his key note address to James's psychology: *William James psychologue: l'intérêt de son oeuvre pour des éducateurs.*[37] Two years later Bovet was the first director of the "Institut Jean-Jacques Rousseau"[38] in Geneva founded by Edouard Claparède, and in 1920, he became professor of education at the University of Geneva. Together with the Genevan sociologist Adolphe Ferrière, Bovet and Claparède dominated the educational scene in Geneva and made it a worldwide center of educational discussion.

What Flournoy came to mean for James, Claparède would—about ten years later—prove to mean for Dewey. The increasing interest in Dewey can be reconstructed by examining one of Edouard Claparède's best-selling books, *Psychologie de l'Enfant et Pédagogie experimentale.*[39] The book grew in content between 1905 and 1922 from 76 to 571 pages, and it was republished after World War II (Claparède died in 1940) in two volumes, from the eleventh edition on containing also a study by Jean Piaget called *La psychologie d'Edouard Claparède.* The book was translated into more than six languages and appeared in English, based on the fourth French edition, in 1911 (*Experimental Pedagogy and the Psychology of the Child*; reprinted 1975).[40]

The original 1905 edition of Claparède's book contains one single reference to James and no references at all to John Dewey, but both authors become more and more important with each new edition published. In 1909 the French journal *L'Éducation*[41] published John Dewey's *The School and the Social Progress.*[42] Two years later, in the introduction to the 1911 fourth edition of Claparède's *Experimental*

Pedagogy and the Psychology of the Child (the basis of the German and English translations), William James is *the* authority for advocating the importance of psychology for education, and John Dewey is *the* authority for having expressed one of the core problems of education, namely, the genetic-functional phenomenon. This replaced the simple, 1905 conception of an "attractive" education. According to the pragmatic idea that life means active adjustment to constraints, the term "functional education" was central in this new edition. In the eighth edition ten years later in 1920, Dewey is additionally *the* authority for child development—based on his texts collected in *The School and Society*. Meanwhile, *L'Éducation* had published in 1912 the translation of *The School and the Life of the Child*[43] and, in 1914, *Waste in Education*.[44]

A 1913 volume containing four Dewey texts published under the title *John Dewey. L'école et l'enfant*,[45] translated by Louis-Samuel Pidoux[46] and edited by Claparède, demonstrates the increasing importance of Dewey in the Genevan discussion. Claparède's introduction to the volume, titled *La pédagogie de John Dewey*,[47] may well be the first elaborated French study on Dewey, and it proves that almost every paper that Dewey had published after 1886 had been read in Geneva. The Dewey quote (cited in English) that heads Claparède's introduction indicates the quite different interest Claparède had in Dewey than Flournoy had in James; it represents the core of what was called in Geneva *education nouvelle* (new education): "Learning?—Certainly, but living primarily, and learning through and in relation to this living."[48]

Claparède's argumentation is interesting. First, he refers to a certain similarity between G. Stanley Hall, who at this time was much more famous in Europe than Dewey, and Dewey, whose theory of education was considered to be "more genetic and dynamic."[49] Listing all the different areas of Dewey's research, Claparède hastens to tell his readers that there has always been just *one* method: "it is *pragmatism*, of which Dewey is, together with William James and F. C. S. Schiller, one of the most brilliant heads." Claparède cites Flournoy's definition of the core of pragmatism, namely, to renounce to the use of decontextualized terms and, instead, to look at real and situated consequences. Claparède demonstrates how Dewey uses this method in his moral philosophy, his logic, and his psychology and mentions that it contains in itself an "educational method."[50]

Most interesting is that Claparède sees in Dewey's "psychopedagogics" a *representation* of pragmatism at its best, but he denies any common destiny with the fortune of pragmatism as a "doctrine," which had been the object of raving attacks in Europe ever since the International Congress of Philosophy in Heidelberg 1908.[51] Claparède's intention is "pedagogic" insofar as he tries to show his deep conviction in progressive education. Dewey's education is, as Claparède says in short, "essentially dynamic" and deeply rooted in life. "Life! Life! Ah! If we want life, let us place ourselves into life; let us see how the child is, and where it is heading."[52]

Dewey's education is, according to Claparède, threefold: genetic, functional, and social. The *genetic* aspect is believed to have overcome the mechanistic and/or analytic-empiric psychology of nineteenth-century Germany by stressing the dynamics of the child's self-development. The focus on the child's development turns out to be very useful for practitioners, because it solves *the* Gordian knot of education: the relation between the arbitrariness of the child and exterior constraints. To leave children on their own to "obey" their desires is no education at all, because the skills and abilities of the child develop only through mastering obstacles. The *functional* aspect reinforces the contrast to the empirical psychology, because it indicates active adaptation to a situation created by interaction of external circumstances and inner desires—and this active adjustment is usually neglected in the traditional schools. And the *social* aspect, according to Claparède, can in fact be reduced to the functional aspect.[53]

Claparède's introduction shows that the three dimensions of education analyzed were estimated unequally. Claparède's own book, *Experimental pedagogy* (from the fourth edition in 1911),[54] proves that the two first elements, "genetic" and "functional," were seen in combination. They are interpreted as one of the core problems of education, while the "social" aspect is hardly discussed at all. This "ignoring" of the social dimension accords with the way in which Dewey's educational theory was viewed in "isolation" from pragmatism. From this, one could easily conclude that the Genevans were not interested in the social and political part of Dewey's philosophy of education, but only in his progressive psycho-pedagogical construction. But why, then, were James and Dewey so much more highly estimated in Geneva than the well-known German progressive educators? The reason for this is to be found not so much in explicit concepts or arguments, but in the same reformist Protestant *langue* that they shared, which made the American foreigners seem indigenous in Geneva.

Reformist and Lutheran Protestantism

It is a remarkable phenomenon that Germany was familiar with James and Dewey at least as early as the French Swiss were. James was known as an important psychologist, and Dewey was known—even earlier than in Geneva—as a reformer of schools: *The School and Social Progress* (1899) appeared in German as early as 1903 and *The School and Society* in 1905. But when it became obvious that James and Dewey were strongly interrelated with pragmatism, sympathy soon waned.[55] James's 1907 volume, *Pragmatism. A New Name for Some Old Ways of Thinking; Popular Lectures of Philosophy*, was published in German only a year later (1908) by Wilhelm Jerusalem. As the discussions during the Third International Congress of Philosophy in Heidelberg 1908 show, arguments in favor of pragmatism had a very hard standing, because it refuted eternal truths and metaphysics. Schiller, for instance, argued that truths were always related to real human life and that the idea of an "independent, supernatural, eternal, incommutable, unachievable, inapplicable, and useless truth" was a childish delusion.[56] Later in the discussion he assessed that this way of looking to truth is useful and leads human kind to a "*progressus in infinitum.*" Truth does not have to be true in the beginning—the likeliness is the utmost certainty we get when assumptions "stand the test of experience."[57] These positions were unique and exposed to attacks, and on this background it is almost an ironic *aperçu* that the Second International Congress for Philosophy had been organized by Flournoy and Claparède in Geneva 1904. They did not attend the Heidelberg congress (neither did James), and Flournoy remarked in a letter to James of September 20, 1907: "I have had news of the Congress at Heidelberg only through [Lorenzo M.] Billia, the philosopher from Turin, who passed through Geneva . . . He found the Congress very tedious, much too German and not international; . . . ."[58]

The scandal within European, and most of all German, philosophy was enormous, and it was in this context that Eduard Spranger, professor of education and philosophy in Berlin, belittled John Dewey's work, which he reduced to education that was merely economic and technical. He assessed it as vastly inferior to the "latitude of German education." For Spranger, Dewey's work represented—in stark contrast to the higher ends woven into the German mind—a despicable kitchen and handyman utilitarianism that had to be countered by the "theory of the ideal *Bildung.*"[59] Compared to this militant refutation of pragmatism in Germany, the "emancipation" of Dewey's education from the destiny of pragmatism, as Claparède had written, was a very careful formulation.

The harsh rejection of pragmatism in Germany does not yet explain why James and Dewey found greater acceptance in Geneva than in Germany, but it indicates a deep difference that I see as religious-cultural. One of James Hayden Tuft's recollections of his year of Ph.D. study at the University of Berlin before coming to the newly founded University of Chicago (where at his urging Dewey and Mead were hired away from Michigan in 1894) throws light on this difference: He describes how when entering the University of Berlin he was required to state his denomination: "I didn't think 'Congregationalist' would mean much in Deutschland. Fortunately a friendly German solved the difficulty by two questions and a syllogism. 'Are you Jewish?' 'No.' 'Are you Roman Catholic?' 'No.' 'Then you *must* be evangelical,' for these are the only possibilities."[60] Evangelical, which Tufts *had* to be, meant German Lutheran Protestantism, which, according to Dewey, was the basis of the German philosophical idealism with its dualistic worldview.[61] The German intellectuals of the time were skeptical about intellectualism and deeply indebted to the dualistic two-world conception of Luther and indebted politically to the German *Reich*, believing in the superiority of the German *Volk*. This mode of thinking necessarily led to deep distrust in conceptions that denied ideas of eternal truths as well as dualisms, that built on interaction and on cooperation, and therefore on democracy. Pragmatism *had* to appear as a danger and was therefore discredited, under the banner of slogans like "dollar philosophy" or similarly disrespectful terms. Swiss reformist Protestantism was based on Zwingli (Zurich) and Calvin (Geneva), and it—transformed by British Protestantism—was dominant in the United States. It is socially and politically very different from German Lutheran Protestantism, both in its original positions of the sixteenth century and in its liberal versions around 1900. Tufts and Dewey were very well aware of that fact; at least it became obvious on the occasion of World War I and again later in the context of World War II.

Langue as Homeland

From this perspective, the Genevans' appreciation of American psychology and their distancing from German psychology becomes more understandable. Claparède's introduction to the French translation of the four Dewey articles admires Dewey's psychology as *against* German psychology, which he characterizes as "static"[62] and criticizes for dividing up development into analytic elements that have no inter-relation to each other[63] and as "sterile doctrine."[64] This critique is very similar to Dewey's objections in his article *Reflex Arc Concept in Psychology* that German psychology, building on the dualism of stim-ulus and response, was too mechanical, "a patchwork of disjointed parts."[65] Dewey's article was first published 1896 in *Psychological Review*, together with articles by colleagues from the philosophy department at the University of Chicago. The articles appeared again in the same year as offprints in the first issue of *Studies from the Psychological Laboratory: The University of Chicago Contributions to Philosophy*, edited by James Rowland Angell[66]—a former student of Dewey at Ann Arbor and son of congregational minister and presi-dent of Ann Arbor, James B. Angell. The anonymous *Introductory Note* (1896) to the issue sets out the purpose of the collection as pre-senting articles that were all based on experiments that were under-lined by a certain hypothesis: "This is the conception of consciousness as an organic unity, with the resulting ideas that the facts are facts of growth or continuous realization, and are to be interpreted as such. It is believed that this principle is as important in its bearing upon labo-ratory work as upon those phases of psychology definitely labeled genetic: that, in other words, mental phenomena are to be regarded as continuous changes in an interaction of organisms and environ-ment. This point of view, of necessity, throws the emphasis upon activity . . . ."[67]

This fundamental way of looking at man's soul was decisive for the positive reception in Geneva, and it explains why James and Dewey, although foreigners, were indigenous. "Organic unity," "growth or continuous realization," "interaction," "activity" were all attractive in psychology, education, and politics framed by liberal Calvinism or secularized Congregationalism. It is no coincidence that George Herbert Mead, the son of a congregational minister, tried to prove in a lecture in Ann Arbor the deep truth of the Sermon of the Mount with the theory of action that he found in James's *Psychology*. To recognize Jesus and therefore the community of interests of all men, Mead writes, would be neither a question of rationality, nor one of "emotion in the sense in which we generally perhaps consider love an emotion. It does not represent a feeling insofar as this is something static but a state [of] mind prepared for the most absolute, the most perfect activity—it is the condition of perfect activity."[68] To prove this statement, Mead cites William James's *Psychology*,[69] where James had insisted that emotions were connected to physical activities and therefore prior to all attempts to rationalize. Mead concludes that the emotion of love occurs only with activity. Love is the emotion stemming from the most complete and most absolute activity of humankind: "I come back to our theme if the principle which Jesus represents is to be expressed as an emotion. The emotion of love—it can only be as an action principle—the principle of the most complete and absolute activity our natures are capable of." It must have "back of it the instinctive" actions of the whole social—in other words: religious—nature and it must have the power of supporting these activities at once "and without cessation."[70]

It is not a coincidence that the two persons in Geneva who were the first to feel "at home" in the pragmatist way of thinking, Flournoy and Claparède, were not only cousins, but that both stemmed originally from French Huguenot families that left France after the revocation of the Edict of Nantes in 1695 by Louis XIV. The Protestant view of a fundamentally active person—as laborer as well as citizen—frames the background of the Genevan interest in pragmatism. Thus we understand why Claparède points out Dewey's "primacy of action" in his introduction in countless variations.[71] The aim of education was "self-realization," which meant calling "all inner activities."[72] According to Claparède, Dewey's psychology built on "mental activity"[73] and on the faculties as "instruments of action," adapting to circumstances.[74] All development is considered to be "genetic,"[75] leading to a "unity,"[76] so that education could not be different than active, for example, manual training. Pierre Bovet, in his William James lecture in 1910, used Herbert Spencer's terms to adapt it to pragmatism. "Life is the continuous adjustment of internal relations to external relations,"[77] and this adjustment, which can be understood as passive (as it still is in Germany today), was interpreted by the Genevans—according to James or Dewey—as *activity*.[78] Whether or not the Genevans had read Dewey's entry on *Adaptation* in Monroe's *Encyclopaedia of Education*, they certainly would have agreed with Dewey's attempt to warn against interpretation influenced by Herbert Spencer. According to Spencer, "adaptation" was reduced to "passive organic beings," an assumption that would lead to a "perversion" in educational thinking, Dewey fears, because it would mean "the accommodation of individuals to the existing type of social polity and customs." "To avoid this error," says Dewey, "it is necessary to realize that adaptation is a case of control involving the subordination of the environment to the life functions of individuals." Adaptation is always *active*[79]—it is the essence of life.

How important "activity" was to the Genevans can be demonstrated not only in the way that they received James or Dewey or in the way that they were "affected" by the American scholars. In addition, it becomes visible in a public conflict between the two colleagues and friends, Ferrière and Claparède. The conflict was about the interpretation of "activity" as the top slogan around 1920: *L'école active*, the *Activity School*. The term had shown up first in 1917 and was in the core of educational semantics within less than eight years. In 1922, Adolphe Ferrière published a two-volume book, *L'école active*, with many references to James and Dewey, and the combined edition of 1929 shows even a significant increase in these references. Almost countless articles in journals used the title *L'école active* to praise the very core of all progressive education. Albert Chessex, a teacher and comrade-in-arms of Ferrière, for instance, demanded in the Genevan journal *L'Educateur* that every student had to be "active in the broadest and highest sense of the term."[80]

The same year, Edouard Claparède published an article in the same journal criticizing the advocates of *L'école active*, but not, of course, because he favored anything like a passive school. He argued first that Ferrière had departed psychology for some metaphysic fields, arguing with unintelligible slogans such as Bergson's "élan vital." The practitioner will not know what to do with slogans like that, and when Chessex contents himself with demanding "a broadest and highest sense" of activity, the practitioner still is left on his own. Then what exactly does Chessex mean? The notion "active" is equivocal, Claparède teaches.[81] According to him, we have to differentiate between two different senses of "active." There is active in the "sense of effectuation" and active in the "sense of function".[82] Claparède advocates for the latter: "Activity is always evoked by needs,"[83] which is his core idea in a book he would publish in 1931, *L'Education fonctionelle*.[84] Activity as a function of adjustment was a Darwinian-Protestant idea that Claparède had seen in Dewey and Rousseau: "If you want to make your child active, put him in circumstances that call for a need to fulfill an action you expect from him."[85] Ferrière, too, when he defended the slogan *L'école active*, saw Rousseau and Dewey as his own forerunner: "If all life is essentially adaptational work, a reaction to the environment, in other words: activity and labor, so we have to admit that Dewey across the ocean was the first pioneer for labor in school, not just for manual labor, but mainly for labor as a whole."[86]

Reception as Activity

The slight relativization of Dewey's pioneer work that one can read in Ferrière's phrase above is no coincidence. It relates to the self-confident Genevans' own ambitions in general, but to Ferrière's in particular. Behind all the recognition of James or Dewey, we find little modesty in the arguments. The Genevans believed that they were fulfilling the program of a "science of education" that had been announced by the eminent French educationalist Gabriel Compayré in 1879 and 1881, urging that history embodies all good ideas of education—French history in the main, of course—and that all that was needed now was the evaluation of those ideas by a pedagogical psychology. This program had been highly attractive to the Genevans,[87] only they switched around the chronology somewhat. They did not evaluate historical ideas with the means of modern psychology (which would have hardly been possible anyhow); instead, they constructed a history of their own pedagogical psychology. Thus, the argumentation architecture of most of the Genevan books was similar. The books usually began with a historical look at comparable enterprises in the past in order to reinforce their own concerns and ideas, and the more renowned the educational "predecessors" were, the better: Rousseau and Dewey were merely ideal witnesses of their own ambitions. Ferrière was just a bit more crass in expressing his pride than his colleagues when he asked Dewey, who had stayed in the French part of Switzerland in 1914, in a letter (in French) whether he was the author of *The School of Society* having been published in French under the title *L'école et l'enfant* (School and the Child).[88] If this were the case, Ferrière writes, he would be delighted to welcome Dewey to Geneva, his own birthplace. And while he assures Dewey that there were "few men with whom I share so completely a philosophical and pedagogical ideal," he names some of his own articles that cited Dewey and invites Dewey to a talk that he would hold in Geneva.[89] A similar strategy is found in his book *L'école active*. After having described the ideas of Dewey, Decroly, and Lighthart (the Dutch "Deweyan"), Ferrière writes: "I, myself, have also—in absolutely autonomous way—taken about the same paths as John Dewey, Decroly, Jan Lighthart, and Lay. Here you find along general lines what I have been doing in practice."[90] How little Ferrière in particular really studied Dewey becomes apparent in his comment on Dewey's *Democracy and Education* in *L'Éducation*. Ferrière holds that *Democracy and Education* had been published only in 1923.[91] Limiting himself to "collecting" the best ideas that Dewey's book contains, Ferrière does little more than point out those elements that support him, for example, his own doctrine. The word "active" or "activity" appears at least fifteen times in his seven-page-long comment. Of course, Ferrière sees in Dewey the "spiritual descendent" of Rousseau as well as *the* exponent of European progressive education "across the ocean." On the other hand, he does not mention any of Dewey's critique of Rousseau, nor do we find any hint of Dewey's idea of active *adaptation* that changes the circumstances, itself causing new constraints. Ferrière ignores the fundamental problem that Dewey fights against, the dualisms in both philosophy and society. When Dewey describes that in democracy the aim of "culture" emanates from education beyond the two false dualistic alternative aims of "social efficiency" and "something purely 'inner'," [92] Ferrière simplifies the problem in his psycho-pedagogical view. He redivides aim and process: "In a democracy . . . we have to cultivate the child's faculty of joining voluntarily and completing the collective activities within the division of labor. This is not possible without a strong culture." [93]

Indeed, the deep confidence in the psychological basis of education and the interpretation of psychology as psycho*logic* eclipsed to a certain extent the social and political dimension of Protestantism, much more so than we find in Dewey. When Ferrière writes that Dewey was "a philosopher rather than a psychologist, but a knowledgeable adept of childhood," [94] he expresses the differences quite well. Dewey's writings about (new) education that were received were closely connected with his early phase in Chicago and his Laboratory School—with the exception of *Schools of Tomorrow*, which he jointly wrote with his daughter Evelyn in 1915. [95] This selection of articles by the Genevans indicates the selection of perception *within* these articles. Dewey's considerations of contingency in life and the danger that democracy faced in a capitalistic society are not really appreciated in Geneva—and a book like Dewey's *The Public and Its Problem* (1927) would have hardly been liked in Geneva.

This basically psycho*logical* fundament of educational thinking steered the reception of pragmatism in Geneva—on the basis of political and psychological reformist Protestantism that they were little aware of. The difference can be seen, for example, in the way that Dewey and Claparède responded to the political situation at the end of the 1930s. Dewey was defending American democracy by referring back to political virtues as portrayed by Thomas Jefferson in *Freedom and Culture* (1938) and the *Living Thoughts of Thomas Jefferson* (1940). Claparède, in contrast, complained shortly before his death of the loss [*vacation*] of individual righteousness in a talk, titled *Morality and Politics*, that he held before the circle called "Friends of Protestant Thinking" in Geneva and La Chaux-de-Fonds (canton of Neuchâtel). Here again, education becomes the means of reinstituting these virtues. But in a naive reading of Rousseau, Claparède says: "Education means to form 'the human,' in other words the person, to lead him to the spiritual unity characterized by the fact that the 'higher self' is capable of perceiving and thwarting the traps stemming from the 'lower self.'"[96] Politically, Claparède was quite at a loss, as can be expected of someone who had a lifelong dedication to interpreting everything according to the psychological point of view and progressive education. He thus shared the *langue* with pragmatism, using it when formulating his *paroles*, but he did not engage in conscious reflective thinking about it.

But this is not the end of the story. An astonishing number of books and articles by the Genevans were translated into other languages, such as Esperanto, Romanian, Polish, English, German, Portuguese, and, by far most of all, in Italian and Spanish. It is obvious that the Genevans were widely read and appreciated, particularly in Catholic countries with rather rudimentary democratic experience (including South America). For example, Edouard Claparède's *Experimental Pedagogy and the Psychology of the Child* was already translated in 1910 into Spanish and reprinted in 1911, 1927, and 1930, and it appeared in Italian in 1934, 1936, 1971, and 1973. His *How to Diagnose the Aptitude of Students* was printed in Spanish in 1924, 1927, 1933, 1959, 1961, 1964, 1967, and 1972. Pierre Bovet's *Feeling of Oughtness*, which appeared in English in 1912, was published in Spanish in 1922, 1925, 1927 (twice), 1928, and 1934. Ferrière's *Activity School*—printed in English 1928—was published in Spanish in 1927 and 1932 and in Italian in 1939, 1947, and 1958; his *Spontaneous Activity of the Child* was translated into Italian only and published there in 1947, 1951, and 1970.

Whether or not the translations of the Genevan books and articles stimulated scholars in the Catholic countries to read primary literature and how this "steered" their active reception of Dewey or James is a question that cannot be answered in this book. "Transfer" is—and here pragmatism proved to be right—unpredictable, because it depends on the interests and activity of the receptors. The principle of activity is not logic—neither psycho*logic* nor teleo*logic*—but contingency. But paradoxically, the international attractiveness of the Genevan publications was based in the psycho*logic* foundation of education that promised to fabricate a "modern self" able to master both the opportunities and challenges of modern industry and democracy. The reduction of the active citizen to the psychological activity of child was in its core still deeply Protestant, but this was difficult to recognize. To most of the educators in the world, who feared the contingent modern developments, these roots remained unrecognizable because of the attractive promise of the comforting vision of well-ordered child development. They were not aware that their own reception was active and contingent, too, and that their conception of the "modern self" could well differ from the vision of the ideal (and idealized) Genevan citizen. The resulting Catholic and (at best) apolitical active reading of Dewey or James or Claparède or Ferrière in Italy or Spain is neither only a sin, nor only degeneration, but perhaps just a bit awkward. But this reception has something to do not only with questions of theoretical compatibility across contexts, but also with the exalted educational semantics that, among others, all four authors were themselves reinforcing around 1900 (and after 1900, too). Maybe it is not a coincidence that the older Dewey got, the less he wrote about education.

Notes

1. I would like to thank Ruth Villiger, Sylvia Bürkler, and Markus Christen, who helped me find the sources.
2. William James, "Are we automata?" *Mind: A Quarterly Review of Psychology and Philosophy* 4, no. 1 (1879): 1–22.
3. William James, "What is an emotion?" *Mind. A Quarterly Review of Psychology and Philosophy* 9, no. 2 (1884): 188–205.
4. Robert C. Le Clair, ed., *The Letters of William James and Théodore Flournoy* (Madison: The University of Wisconsin Press, 1966), xiiiff.
5. William James, *Principles in Psychology* (New York: Holt, 1890).
6. Le Clair, *The Letters of James and Flournoy*, 6.
7. Flournoy's booklet resulted out of studies on the relation between the soul and the body. His aim is to develop a psychology that has "the character of a science" Théodore Flournoy, *Métaphysique et psychologie* (Genève: Georg, 1890), V. The bases of this psychology are first of all the "laws" behind the so-called psychophysiological parallel and, second, the effort to emancipate psychology from the field of metaphysical philosophy. Flournoy legitimated this effort by the "empiric fact" that "psychophysiological parallelism" was dualistic. According to tests in his laboratory, there was no mutual relation between psychological facts and physiological events—they are just "simultaneous" (17). From the point of view of an experimental science, the psychophysical parallel is a given thing, inexplicable, an enigma, that seduces people into metaphysics (20). Flournoy insists that all metaphysical explanations and systems are fruitless attempts (51).

8. Le Clair, *The Letters of James and Flournoy*, 7f. A photograph of Flournoy and James in Geneva on May 18, 1905 is available on the Internet at: http://www.emory.edu/EDUCATION/mfp/jamesn.html.

9. "My dear Flournoy, there is hardly a human being with whom I feel as much sympathy of aims and character, or feel as much 'at home' as I do with you" (August 30, 1896); Le Clair, *The Letters of James and Flournoy*, 54.

10. The notion of republican *langue* does not refer to contemporary political parties in the Western world, be it in France, Germany, or the United States. Republicanism was a political mode of thinking, a language that was dominant in specific epochs of the Roman Empire, in some of the late medieval cities in Italy such as Florence, in Switzerland since its foundations, in England during the Commonwealth in the middle of the seventeenth century, and in the United States since the discussions in 1776. It is usually an anticapitalistic and pro-agrarian political ideal, as we find, e.g., in the writings of Thomas Jefferson. Liberty as the citizens' right to self-government is the core of this way of thinking, including the virtues that these rights presuppose: unselfishness, non-corruption, love of liberty, and love of the law. A (broadly discussed and often challenged) history of republicanism from Florence to the discussions of independence is provided by Pocock, see: J. G. A. Pocock, *The Machiavellian Moment. Florentine Political Thought and the Atlantic Republican Tradition* (Princeton: Princeton University Press, 1975); an analysis of the discussions in and after 1776 in the United States by Woods, see Gordon S. Woods, *The Creation of the American republic 1776–1787* (Chapel Hill: University of North Carolina Press, 1969). Although this political mode of thinking builds on virtue(s), republicanism has hardly been discussed in education, regardless of the fact that Rousseau and Pestalozzi were committed republican citizens: The dominant historiography of education has successfully suppressed this tradition, see Daniel Tröhler, "The establishment of the standard history of philosophy of education and suppressed traditions of education," *Studies in Philosophy of Education* 23, no. 5/6 (2004), 367–391. The influence of the mixture of Protestant and republican thinking in the United States in the nineteenth and twentieth centuries on school politics is described in David Tyack and Elisabeth Hansot, *Managers of Virtue. Public School Leadership in America, 1820–1980* (New York: Basic Books, 1982).

11. Le Clair, *The Letters of James and Flournoy*.

12. John Dewey, "The reflex arc concept in psychology," in *Studies from the Psychological Laboratory: The University of Chicago Contributions to Philosophy* 1, no. 1, ed. James R. Angell (Chicago: Chicago University Press, 1898).

13. Only the human being / is able to do the impossible. / He distinguishes, chooses, and judges, / and he can give permanence / to the moment (freely translated here), See, William James, *Are we automata?* 15.

14. Dorothy Ross, *The Origins of American Social Science* (Cambridge: Cambridge University Press, 1991); George M. Marsden and Bradley J. Longfield, *The Secularization of the Academy* (New York: Oxford University Press, 1992).

15. Louis Menand, *The Metaphysical Club. A Story of Ideas in America* (New York: Farrar, Straus und Giroux, 2001), 141.

16. William James, *The will to believe, and other essays in popular philosophy* (New York: Longmans, 1897), 319.

17. Axel Munthe, *Story of San Michele* (New York: Dutton, 1929), 372–373.

18. William James, *Varieties in Religious Experience* (New York: Longmans, 1902).

19. William James, *L'expérience religieuse: essai de psychologie descriptive*, trans. Frank Abauzit (Genève: Kündig, 1906).

20. Le Clair, *The Letters of James and Flournoy*, 213.

21. Théodore Flournoy, "Les principes de la psychologie religieuse," *Archives de Psychologie* 2, no. 1 (1903): 43.

22. Théodore Flournoy, "Les principes," 38.

23. This institute is now a part of UNESCO. See: http://www.ibe.unesco.org/AboutIBE/hise.htm.

24. Pierre Bovet, *Le sentiment religieux et la psychologie de l'enfant* (Neuchâtel: Delachaux & Niestlé, 1925).

25. William James, *Pragmatism. A New Name for Some Old Ways of Thinking; Popular Lectures of Philosophy* (New York: Longmans, 1907).

26. "I am all aflame with it [Pragmatism, DT]," and "I want to make you all enthusiastic converts to 'pragmatism' " (James to Flournoy, January 2, 1907; Le Clair, *The Letters of James and Flournoy*, 181).

27. Le Clair, *The Letters of James and Flournoy*, 193–194.

28. John Dewey, *Studies in Logical Theory* (Chicago: University of Chicago Press, 1903).

29. Dewey's book, by the way, was dedicated to William James, who was not insusceptible to vanity and reacted quickly by publishing a short article, *The Chicago School*, in *Psychological Bulletin* in 1904, see William James, "The Chicago School," *Psychological Bulletin* 1 (1904):1–5. James praised Dewey as the head of this innovative school, just as Dewey was about to leave Chicago for Columbia, see Le Clair, *The Letters of James and Flournoy*, 152.

30. Le Clair, *The Letters of James and Flournoy*, 187.

31. Le Clair, *The Letters of James and Flournoy*, 205.

32. Le Clair, *The Letters of James and Flournoy*, 217.

33. A German translation had been published in Germany already in 1900, and it was published in Italian as early as 1903. See William James, *Talks to teachers on Psychology and to Students on some life ideals* (New York: Longmans, 1899); William James, *Causeries pédagogiques* (Lausanne: Payot, 1907).

34. H. Blaser, "Les Causeries pédagogiques de William James," *Bulletin mensuel du Département de l'Instruction publique* 9, no. 3 (1907): 21–25.

35. M. Métral, "Résumé des 'Causeries pédagogiques' de W. James," *L'Éducateur* 43, no. 16 (1907): 260–261.

36. Louis[-Samuel], Pidoux, "Le philosophe William James," *L'Éducateur* 43, no. 31–32 (1907): 485–487.

37. Pierre Bovet, "William James psychologue—L'intérêt de son œuvre pour des éducateurs," *Bulletin mensuel du Département de l'Instruction publique* 12, no. 10 (1907): 115–132.

38. Today the institute is a part of the University of Geneva. See: http://www. unige.ch/rousseau/welcome.html.

39. Edouard Claparède, *Psychologie de l'Enfant et Pédagogie experimentale* (Genève: Kündig, 1905).

40. The very same fourth French edition was the basis of the German translation in 1911. See Edouard Claparède, *Psychologie de l'Enfant et Pédagogie experimentale*, 4th ed. (Genève: Kündig, 1911); Adolphe Ferrière, *L'école active*, Vols. 1 and 2 (Genève: Forum, 1922).

41. The editor of this journal, Georges Bertier, was the director of the experimental school *L'École des Roches*, and in the same year he translated William James's *Principles*. He advocated a secular state and was active in the Protestant dominated French Boy Scouts movement that today call itself *l'Unité scoute protestante Jean Calvin* (Protestant Scouts Unity Jean Calvin).

42. John Dewey, "L'école et le progrès social," *L'Education* 1: 198–217.

43. John Dewey, "L'école et la vie de l'enfant," *L'Education* 3: 315–327.

44. John Dewey, "Le gaspillage en éducation," *L'Education* 6: 8–25

45. These were: *L'intérêt et l'effort* (Interest and Effort in Education, 1913); *L'enfant et les programmes d'études* (The Child and the Curriculum, 1902); *Le but de l'histoire dans l'instruction primaire* (The Aim of History in Elementary Education, 1902); *Morale et éducation* (Moral Principles in Education, 1909).

46. Louis-Samuel Pidoux had already translated William James's *Will to believe* under the title *La volonté de croire*, which was edited in Saint-Blaise near Neuchâtel 1908, and *Talks to teachers on psychology and to students on some life ideals*, which was edited in 1907 in Lausanne and published under the title *Causeries pédagogiques* (see Louis[-Samuel] Pidoux, Le philosophe).

47. Edouard Claparède, "La pédagogie de John Dewey," in *John Dewey. L'école et l'enfant* (Neuchâtel: Delachaux & Niestlé 1913).

48. Claparède, "La pédagogie de John Dewey," 5.

49. Claparède, "La pédagogie de John Dewey," 6.

50. Claparède, "La pédagogie de John Dewey," 13.

51. Theodor Elsenhans, ed., *Bericht über den 3. Internationalen Kongress für Philosophie zu Heidelberg, 1.–5. Sept. 1908* (Heidelberg: Winter, 1908).

52. Claparède, "La pédagogie de John Dewey," 14–15.

53. Claparède, "La pédagogie de John Dewey," 33.

54. Claparède, *Psychologie de l'Enfant*.

55. Philipp Gonon, "Dewey und James in Deutschland—Verpasste Rezeptionschancen des amerikanischen Pragmatismus," in der deutschen Pädagogik. *Pragmatismus und Pädagogik. Gesellschaftstheorie und die Entwicklung der (Sozial-)Pädagogik*, ed. Daniel Tröhler and Jürgen Oelkers (Zurich: Pestalozzianum, 2005).

56. F. C. S. Schiller, Der rationalistische Wahrheitsbegriff. *Bericht über den 3. Internationalen Kongress für Philosophie zu Heidelberg, 1.–5. Sept. 1908* (Heidelberg: Winter, 1909), 711.

57. F. C. S. Schiller, Schlusswort. *Bericht über den 3. Internationalen Kongress für Philosophie zu Heidelberg, 1.–5. Sept. 1908* (Heidelberg: Winter, 1901), 739.

58. Le Clair, *The Letters of James and Flournoy*, 202. One of the most striking examples of foreignness between the Germans and pragmatism can be found in a rather unknown text, namely, in the introduction to the German translation of the book *La philosphie de William James*, which was published by Flournoy a year after James death in 1910 in French (in 1917, it appeared in English translation in New York); (See Théodore Flournoy, *The Philosophy of William James*, trans. Edwin B. Holt and William James, Jr. [New York: Holt, 1917]). In this introduction, the German editor Arthur Baumgarten, who had studied under Flournoy in Geneva, praises James as the greatest philosopher after Schopenhauer—but with the severe blemish of a "pragmatistic theory of truth": "Verifying through experience and practical probation are most worthy indications for the truth of an assumption, but it is not acceptable to identify it with truth." See Arthur Baumgarten, "Einführung," in *Théodore Flournoy, Die Philosophie von William James* (Tübingen: J. C. B. Mohr, 1930), x. Also see: Théodore Flournoy, *La philosophie de William James* (Saint Blaise: Foyer Solidariste, 1911); Théodore Flournoy, *Die Philosophie von William James*.

59. Daniel Tröhler, "The Discourse of German *Geisteswissenschaftliche Pädagogik*—A Contextual Reconstruction," *Paedagogica Historica. International Journal of the History of Education* 21 (2003): 759–778.

60. James Hayden Tufts, Germany. University of Chicago, Regenstein Library, Special Collections, The James H. Tufts Papers, Box 3, Folder 13, n.d.

61. John Dewey, *German Philosophy and Politics* (New York: Holt, 1915).

62. Claparède, "La pédagogie de John Dewey," 11.

63. Claparède, "La pédagogie de John Dewey," 17.

64. Claparède, "La pédagogie de John Dewey," 21.

65. John Dewey, "The reflex arc concept in psychology," 39–40.

66. James Rowland Angell, ed. *Studies from the Psychological Laboratory: The University of Chicago Contributions to Philosophy* 1, no. 1 (Chicago: Chicago University Press, 1896).

67. James Rowland Angell, ed., "Introductory Note," in *Studies from the Psychological Laboratory*.

68. George Herbert Mead, [Untitled essay on Jesus, love, and activity] in University of Chicago, Regenstein Library, Special Collections, The Mead Papers Box 10, Folder 1, 26.

69. William James, *Principles in psychology*, 449–450.

70. George Herbert Mead, [Untitled essay], 37–38. Punctuation: DT; asterisk = uncertain reading of Mead's handwriting.

71. Claparède, "La pédagogie de John Dewey," 13.

72. Claparède, "La pédagogie de John Dewey," 15.

73. Claparède, "La pédagogie de John Dewey," 20.

74. Claparède, "La pédagogie de John Dewey," 21.

75. Claparède, "La pédagogie de John Dewey," 16.

76. Claparède, "La pédagogie de John Dewey," 24.

77. Pierre Bovet, "William James psychologue," 3.

78. Daniel Hameline, "Profil d'Éducateurs: Édourad Claparède," *Perspectives* 26, no. 3 (1986): 437–444.

79. John, Dewey, "Adaptation," in *A Cyclopedia of Education*, ed. Paul Monroe. Vo1. 1, no. 35 (New York: The Macmillan Company, 1911), 35.

80. Albert Chessex, "A propos de l'école active," *L'Éducateur 59*, no. 15 (1923): 241–245.

81. Edouard Claparède, "La psychologie de l'école active," *L'Éducateur 59*, no. 23 (1923): 371.

82. Claparède, "La psychologie," 377.

83. Claparède, "La psychologie," 372.

84. Claparède, *L'éducation fonctionelle* (Neuchâtel: Delachaux & Niestlé, 1931).

85. Claparède, "La psychologie," 373.

86. Adolphe Ferrière, *Schule der Selbstbetätigung oder Tatschule* (Weimar: Böhlau, 1928).

87. Nanine Charbonnel, "La pensée pédagogique de Gabriel Compayré: ou la longue marche (de l'éducation) vers le moderne," in *Die neue Erziehung. Beiträge zur Internationalität der Reformpädagogik*, ed. Jürgen Oelkers and Fritz Osterwalder (Bern: Peter Lang, 1999).

88. Dewey, "L'école et la vie de l'enfant."

89. The letter is not only embarrassing because Ferrière is obviously trying to display to Dewey his own ingenuity, but also because the articles collected in Claparède's *L'école et l'enfant* were not identical with the articles in *The School and Society* (see earlier), Adolphe Ferrière, Letter to John Dewey, February 5, 1914, *Past Masters. The Correspondence John Dewey* (Charlottesville, VA: InteLex Corporation, 1914).

90. Adolphe Ferrière, *Schule der Selbstbetätigung*, 226.

91. Adolphe Ferrière, "La démocratie et l'éducation selon John Dewey," *L'Education 18*, no. 5 (1927): 274.

92. John Dewey, *Democracy and Education* (New York: The Free Press, 1916/1944).

93. Adolphe Ferrière, "La démocratie," 276.

94. Adolphe Ferrière, "La démocratie," 274.

95. Dewey, "Letter to Evelyn Dewey, August 4, 1913," *Past Masters*.

96. Edouard Claparède, *Morale et politique ou les vacances de la probité* (Neuchâtel: Edition de la Baconnière, 1940), 178.

4

Dewey in Belgium: A Libation for Modernity?

Coping with His Presence and Possible Influence

Tom De Coster, Marc Depaepe, Frank Simon, and Angelo Van Gorp

The Libation Hypothesis as a Point of Departure

Frans De Hovre (1884–1956), the (anglophile) pioneer of Catholic pedagogical theory in Flanders, repeatedly[1] attributed to the American philosopher and educator John Dewey (1859–1952) worldwide renown as well as an impressive impact on the educational world of the time. But did this presentation of things by De Hovre ever actually correspond to the reality? When we summarily consider the history of Belgian education, we observe that any great presence of the non-Catholic Dewey is probably not so very obvious.[2] Even the Catholic primary school teacher Edward Peeters had to take refuge in pseudonyms and was forced to resign because his interest in the New Education allegedly gave non-Catholic modernists too loud a voice.[3]

Nevertheless, the same New Education gradually became part of the general pedagogical discourse that integrated to a large extent the different views of the actors in the educational field, whatever their political and/or ideological backgrounds might have been. In our view, this discourse developed in Belgium as an educationalized answer to the needs of "modern" society as some of us have been able to demonstrate elsewhere.[4] Catholics and non-Catholics (mainly Liberals and, later on, also Socialists) did not differ in the long run too much from each other concerning the educational conceptions that had to underpin their struggle for institutional power in the schools. The first official curriculum for the primary school dates from 1880, fifty years after the so-called Belgium Revolution, in which the country gained its independence. It was of a Liberal cast and honored unity, simplicity, clarity, and universal applicability as the underlying didactic principles. The Catholics did not want to challenge these principles, not even when they returned to power. What counted for both groups, was, apart from the possible tensions and conflicts about the teaching of religion, the moral character of the school, in which the Socratic maxim of the Enlightenment, knowledge is virtue, had to be applied. The introduction of the working-class children to their place and role in society took place in response to the imperatives of bourgeois society—a social order that rested on a class difference; whether it was created by God or not did not matter. In such a context the formation of educational theories (and science[s]) that would occur from the late nineteenth century onward mainly had a legitimizing function.[5] For example: the Liberal program of 1881 for the "normal schools" (the teacher training of primary school teachers) provided for a discussion of the masters of the "newer" pedagogy, like Montaigne, Comenius, Locke, Fénelon, Rousseau, Pestalozzi, and Fröbel. Much more important than the content element of this educational canon was the prospect of incalculating a "pedagogically" correct attitude into future teachers.[6] This attitude did not change very much during the twentieth century. As long as the school remained the preferred locus of education, a monopoly that was not even questioned when one sought to integretate New Education ideas in the 1930s, the responsibility for its success or failure rested undiminished on the professional competence of the teacher in pedagogy. The primary school Curriculum of 1936, which was considered allover the world as being in the strand of the New Education, stipulated "in addition to the essential science, there is the wonderful art of education." Discipline on the basis of the nature of the child envisioned an orderly school as well as an orderly society in which there was place for authoritative leadership. Like no other, the curriculum designers praised the "prudence," the "wisdom," the "pedagogical tact"—all concepts that had developed in the educational ideology of the end of the nineteenth century—and they realized that it was easier to formulate these principles than to apply them.

In our view, research about the way in which "Dewey"—as far he was and is known in Belgium—was used, and possibly abused, must be framed within this context. As an icon of the New Education, he served, according to us, as a libation for modernity, that is, as a libation for a modern school in a modern society. In such a society "schoolish" life was, in preparation as well as while waiting for "real" life, diluted to a prestructured entity of well-organized relations and networks, whereby it gradually took on more and more the character of an educational island in society, on which pupils were isolated during a moratorium period. They were forbidden to take part in "real" life. On the other hand, the society itself became, as some of us have argued in *Order in Progress* more and more "schoolish." Not only because the role of the school steadily expanded in the society but also primarily because the late-capitalist and social-democratic social ordering appeared to want to appropriate itself constantly the schoolish characteristics of order and authority.

However, before discussing the "context" of the use, and possible abuse of Dewey in Belgium we do need empirical evidence (i.e., the "text") of his "presence" and possible "influence." Such studies do not exist at the moment, nor does, in our view, an adequate and appropriate methodology to undertake such research.

Operationalizing Research Questions about Presence and Influence of Dewey into the Belgian Context

The concept of "presence" has, at least, a double meaning and can be linked to what traditionally is indicated within the German research culture as *Rezeptionsgeschichte*. First, it concerns the material presence of Dewey's progressive educational heritage. How were his publications and translations actually used?[7] Second, the term refers to the reception of Dewey within the educational canon as this is formulated, for example, in encyclopedias and in the educational press. The notion of "influence," on the other hand, focuses on the incorporation of Dewey in the educational world. It is a much used but little defined and even less operationalized category in educational historiography. Moreover, we cannot concur with the general presentation of "influence" in the history of education circles as a "chain of ideas" that, apparently, runs from one author to another.[8] Indeed, an idea does not arise in a vacuum.

Therefore we will also try to examine the incorporation of Dewey's ideas from the framework of *Wirkungsgeschichte*. Such an approach proceeds from clusters and networks developing a pedagogical creed, part of a universal pedagogical canon, "common speech" and "mutual interests"—the general pedagogical discourse we had in mind in the introduction. In other words, in combination with mixtures and assimilations with other factors and ideas, in the nature of the Belgian context, a compilation developed that can be deemed to be "melting pot." The inclusion of Dewey in this (whether or not identified by the mention of his name in the literature), is, of course, more difficult to discern than materially demonstrable presence. However, the study of implicit assimilation traditionally constitutes the core domain of this *Wirkungsgeschichte*.

Belgium certainly is an interesting case for such a study.[9] First of all, it is multilingual, an element that, at first glance, can play an important role in the search for Dewey's ideas or heritage.[10] In Belgium, we can distinguish at least two "cultures": the Flemish (i.e., Dutch speaking) on the one hand and the Walloon (i.e., French speaking) on the other. The historical dynamics of the two—in the sense of an uncoupling process from the unitary Belgium—has been traditionally seen from the Flemish standpoint as an emancipation process. Since this process, after a long history, has become a reality only since the 1970s, a division would be to no purpose, since Belgium had previously profiled itself predominantly as a unitary state. However, we will indicate, where appropriate, the individual accents given by the constituent regions. In addition, Belgium is situated within Europe at the crossroads of several cultures, and—by the multilinguality of the academically educated—many had access to these cultures (the most obvious being the French, the English, the Dutch, and the German). Consequently, Belgium had and has no idea-historical frontiers but was and is taken up in a "traveling library of ideas," also and mainly in the area of education. The relativity of geographical frontiers can also be demonstrated because the promoters of the heritage of the New Education and Dewey in their turn having again hastened also in Belgium the international transfer. Belgian opinion formers had an effect also beyond the borders of their small kingdom.[11] Restriction to Belgian publications would be senseless since foreign books were often used in educational higher education. Thus, the Catholic Dutchman, Frater Sigebertus Rombouts—who, for his work itself, was also greatly indebted to De Hovre—acquired considerable renown in Flanders for such things as his handbooks for the history of education.[12]

But to return to the Belgian situation, one also has to take into account the political and ideological differences between the Flemish and Walloon cultures. Flanders had traditionally a predominantly Catholic cast, which manifested a privileged connection with denominational education. Wallonia, however, underwent the process of secularization much sooner. It has a primarily non-Catholic majority that is emphatically influenced by Socialism and is associated with public education. Thus, we can understand why the protagonists of the educational mentality in Flanders must be situated not exclusively but still in large measure within the Catholic world, with De Hovre, for the first generation, and D'Espallier, a teacher and inspector that became during the interwar period a university professor of education in Leuven, for the second period, as exponents. Next to them, of course, were also representatives of public education (Verheyen, who had a similar career as D'Espallier, at the State University of Ghent for example), authors whom we will treat here as a kind of control group as regards the reception of Dewey in order to determine the extent to which their adaptation deviated from that of the Catholic version. In Wallonia and the French-speaking part of Brussels, where educational notions had received a rather progressive-liberal coloring from the past, the main players were ideologically Masons (like Decroly, who was politically a Liberal) and politically Socialists (Coulon, Clausse). Frère Léon and Frère Anselme were their Catholic opponents in the French-language area—but perhaps one might speak here rather of parrots.

On the basis of these concepts of "presence" and "influence," we wonder about the way in which the activeness—more by implicit than by explicit assimilation—of "reception and assimilation processes" precisely functions. Protagonists feverishly sought in Dewey's works, as they did with ideas they found in other great thinkers and opinion makers at home and abroad, elements to defend their "Catholic" or their "official"-pedagogical practice.[13] By pedagogical-didactic processes of adaptation, filling in, discoloring, didacticization, and simplification,[14] their practice could be legitimized. Indeed, not everyone who wrote about Dewey had, in fact, read him, let alone understood him. The result in Belgium is that Dewey was taken up as an *indigenous foreigner*, a foreigner who was part of one's own heritage, in order to serve "the higher purpose." This finality of pedagogy in general and of the educational reform following the New Education principles in particular, is the service for the salvation of the child.

Within the meritocratic project of modernity, the increased pedagogical action, however, still had to assure "order in the progress."

We may ask, for example, the question of what remains of the complexity of Dewey's ideas when they have to be learned by rote by the "consumers" of the learning process. Thus, pedagogical thinkers were, it is true, traditionally on the program of teacher training and inspector's examinations, but the course in the history of education—in which they were treated—was part of the general pedagogical discourse that we outlined above. Further, teachers and educators had handbooks, but their authors excelled rather in the art of copying. In the best case, these books are constructed from very summary syntheses on the basis of the author being treated or from data gleaned from pedagogical opinion makers. These syntheses, too, were always further reduced, simplified, and didacticized. Thus, for example, we can point to the severely truncated image of Dewey in *Paedagogische Denkers*.[15] As an exponent of the American society, he is painted as the man of the "work school"—naturalist, pragmatist, and instrumentalist—who gave too much attention to sociology and too little to religion.

Of course, on the basis of the limited number of sources we have consulted for this, no detailed picture can be sketched of the filtering and dissemination processes that operated on the heritage of Dewey in our country. What follows is thus only an initial and very rudimentary attempt to frame all this better in preparation for future research rather than to map in an encyclopedic manner Dewey's after-effect. For this, the preliminary research is completely lacking.[16] Indeed, the processes we have in mind, precisely because of the problem of sources, are difficult to study. Therefore, we focus primarily on the heuristic element and try to mark paths for an appropriate analysis of the problem of reception and *Wirkungsgeschichte*. For this, we can use such things as the interchange between "higher" and "lower" pedagogy, layers that we have already introduced previously as ordering categories of the pedagogical mentality.[17] For us the pedagogical mentality is first of all constructed at the basis, by educational agents, like authors of educational journals, who set out what education actually means in practice, thereby limiting the possibility for action. This layer can be called "lower pedagogy." In constructing this lower pedagogy, educational agents constantly reiterate the principles of "higher pedagogy," that is, higher discourse of education that is mostly prophetic. It relates to the "inspiring" ideas of predecessors (Rousseau, Pestalozzi, Dewey, etc.) and the academic pedagogy that perpetuates this tradition (humanities) or attempts (empirically or otherwise) to react against it. Here, we will try to determine how the Deweyan "progressive" heritage in "higher" pedagogy was applied as theoretical legitimization of what was being done in educational practice. Then, we will focus on the uses of his "progressive" heritage in "lower" pedagogy (the educational mentality that governed the dissemination of all kinds of so-called child-centered innovations). And finally, we will try to say something about the filtering processes between these two layers of pedagogical mentality. In sum, attention will be given to the "generational" influence in the dealing with the heritage from the pedagogical canon, of which Dewey certainly was a part. Indeed, in this, the transmission from teacher to student played a decisive role.

Works, Translations, and Presence in the Pedagogical Canon

Those who would want to investigate the presence of works and translations of Dewey can, first of all, focus on the academic world. This assumes an orientation to the personal possessions of professors of pedagogy. What did their libraries look like? Which of Dewey's works did they contain? What happened to them later on? It is possible that the collection was passed on to "successors"—but, as will be seen further on, that was not obvious—or that it was donated to the library of the university where the academics worked. This last possibility would seem to be easier to deal with by the researcher than the generally catalogued but very disparate archives of professors of pedagogy, of whom in Belgium there is not even a prosopography.[18] Do these institutions have works by Dewey (and, if so, from when)? Who purchased them or were they donated? To such questions, one can try to obtain an answer from computerized or handwritten cards. Aware of the methodological pitfalls that such work involves (as well as the patience of a saint that would be required), we intend here to indicate only a few tracks that are relevant to the problem.

As an example, we refer to the Central Library of the *Katholieke Universiteit Leuven*, where we could find the most systematic and complete information. Needless to say, that this institution, till 1968 a unified castle of academic training of future Catholics, played an important role in the formation of the Belgian intelligence. The purchases, gifts, and registers, however, date from after World War II, a restriction that is the direct result of the burning of the library in

1940. Card files in combination with registration lists and computerized data showed that several works by Dewey found their way to the Leuven library. A few purchases by university professors such as the pedagogue Albert Kriekemans cannot hide the fact that the lion's share of the works of Dewey were donated to the library. First, there were the donations from America, which, after World War II, set out to hasten the reconstruction of Western Europe in general and the Leuven university library in particular. Second, a considerable portion of the Dewey collection consists of the donation of the personal library of Canon De Hovre (in 1955–1956) to the university. In addition, the donation is striking of the Leuven professor of experimental pedagogy, Raymond Buyse (four books, in 1977, 1994, and 1997). Consequently, we can say that what the library has is not the result of systematic purchasing but is rather dependent on generous donors who had an eye for Dewey.

As for the presence of translations, re-editions, and finally also for the entire collected works, we note that there is absolutely no "Flemish" translation. The *Paedagogische Encyclopaedie*, the *Standaard Encyclopedie*, and the handbook of Cyriel De Keyser[19] mention only the Dutch translation of *The School and Society* (1929) by T. J. de Boer in the Netherlands.[20] After that, it was not until 1999 before another work by Dewey, *Experience and Education* by Gert Biesta and Siebren Miedema[21] would be published, again in the Netherlands. The connection with the Dutch is important for us because it was precisely these works that constitute the corpus of the Dutch-language references to Dewey.[22] Geographic borders thus form no adequate limitation, and linguistic borders also have to be taken into account. This implies that we, if we are to sketch a precise picture of Dewey's presence in Belgium, also have to take the situation in the Netherlands into account.[23]

The appearance of the collected works[24] was undoubtedly a catalyst for the further dissemination of Dewey's publications. In the year 2004, any self-respecting scientific library is unimaginable without their multi-color bindings. Before the publication of his *opera omnia*, the situation was different in Belgium. Then, only a handful of Dewey's writings were available, certainly as regards to his pedagogical works. In addition to the study of the attention that this extensive work yielded, his death on June 1, 1952 can also be approached as a possibility for the investigation of the peaks in the interest for his heritage. Obituaries can do good service in this regard for obtaining more insight into this attention, but again we can only offer this track as a possible route for further research.

The limited number of mentions of Dewey—only twenty-two—in the index of names of the *Bijdragen tot de geschiedenis van het pedagogisch leven in België*, which provides a detailed overview of the pedagogical press in Belgium up to 1940,[25] shows that the reception of Dewey never scored at the top in the pedagogical periodicals. The interbellum period accounts for most of these references, fifteen in total. Depending on the orientation of the journal—for or against Dewey—he was either listed in a row of reformers or great educators or under the "worshipers" of experimental science, who, through their "idolatry" had lost sight of the true God.[26] Sometimes, however, Dewey received a presentation of his life and works out of interest for renewal, in particular for the so-called active methods or *l'école active*.[27] We refrain, however, from risky conclusions and speak here only of an initial indication. Indeed, staff members varied and the editors of the *Bijdragen* had made no precise agreements about the approach of the articles. Therefore, the journals themselves must be examined.

On the basis of an introduction via the pedagogical encyclopedias, moreover, we note that this limited supply—which is not surprising— virtually coincided with what was seen as the "hard core" of his theory of education. Both the *Standaard Encyclopedie* (1974), and the *Paedagogische Encyclopaedie* (Volume I in 1939, Volume II in 1949) as well as the *Katholieke Encyclopaedie* (1951) mention *The School and Society, How We Think, Schools of Tomorrow*, and *Democracy and Education* as Dewey's standard works. The syllabi, course texts, and textbooks of prominent Catholic pedagogues like De Hovre or, for example, Leuven professors in pedagogy like Nauwelaerts and De Keyser, end up each time with the same works.[28] In 1925, *How We Think* became a permanent feature as obligatory examination material for inspectors of primary education.[29] Nauwelaerts also called attention to *My Pedagogic Creed* and *Experience and Education* by citations in his course text.[30]

Methods for Examining Influence

Where *Rezeptionsgeschichte* can make relatively easy use of search strategies with the name "Dewey" in, for example, encyclopedias, libraries, or the pedagogical press—which will be facilitated once all these files are electronically accessible—searching in the framework of *Wirkungsgeschichte* becomes considerably more difficult. Classic source research with regard to such questions proceeds from pedagogical brochures, press, normal school textbooks, handbooks in the series from the *Vlaams Opvoedkundig Tijdschrift*, and so on. But actually, the study of these sources leaves the researcher with an unsatisfying answer to his or her question. How on earth can one find the "influence" of Dewey if it has already merged into a larger pedagogical canon or if his basic ideas have already been processed by others and are attributed to them? Perhaps, by studying the biographies and libraries on the educational opinion makers.

Among the most important disseminators of the pedagogical ideology are the inspectors, who were often former teachers. For this reason, their notes and markings in the works through which they came to know the pedagogical heroes are very interesting.[31] That these markings are then generally made in schoolmasterly red and blue—the famous two-color pencil served for a couple of decades as an icon of the "corps"—makes this hunt for the dealing with the pedagogical heritage of the "great thinkers" all the more fascinating. A telling example can be found in the copy of the *Paedagogische Encyclopedie* that was worked over by the school inspector, Michel Steels. Underlined passages that also received markings in the margin spring immediately to the eye, and give an impression of the slogans by which members of the profession came to be acquainted with Decroly and of the message that was retained.[32] Such assimilation generates the idea that Dewey's thoughts must have been very straightforward, consistent, and simple. It would therefore be wise to query such filters more systematically in future.

In this same regard, one can, of course, also consider more explicit, personal testimonies that were written down in letters, memoirs, diaries, interviews, and so on. Particularly appetizing is, for example, the travel report of Raymond Buyse, which is preserved in the Leuven library, about his trip with "Master" Decroly to the United States in 1922. In April of that year, the two Belgian fellows of the Commission for Relief stayed a few days in New York at Teachers College where Dewey was a professor at that time. There they met Dewey very briefly: "At 3 o'clock we [Buyse and Decroly] went to Dewey's office to say hello to one of the biggest philosophers in the world. The mood was cheerful. He was a bit shy, sitting there between a pile of books."[33] It was probably a typical situation, but it also told something about Dewey's position regarding his Belgian fans (because that was what they were). Say "hello," a brief conversation out of politeness, and "goodbye." Decroly and Buyse wanted to talk about educational methods with Dewey, whom they called "the Master," but he didn't have enough time. Friendly, but firmly he referred his fans to the Park

School. There, they would find what they were looking for![34] But prudence is called for in the interpretation of the reaction of Dewey. Elements of which the Belgians were not aware could have been involved. For example, did Dewey see them as practitioners with whom a discussion "on an academic level" was not necessary? Or was his interest in 1922 actually focused more on cultural topics? Or were his thoughts in China, where he had traveled for two years (1919–1921)? On the basis of answers to these hypotheses, the filtering into the published scientific travel report can then be investigated.[35]

In our hunt for "influence," we deal with three categories: clearly demonstrable traces in the heritage, the sliding of Dewey into the pedagogical canon, and, finally, the penetration of Dewey.

Clearly Demonstrable Traces in the Heritage

For the directly demonstrable traces of Dewey's heritage, we can best start from the analogy with Ovide Decroly (1871–1932), who, for that matter, is sometimes called the "European Dewey" in the United States.[36] Decroly was a neurologist, who became a world-famous Belgian educationalist, a "pioneer" in Progressive Education and a leading figure in the New Education Fellowship, in particular in the French-speaking branch of the Fellowship. After World War I, in the late 1920s, he also had been appointed a professorship in psychology at the *Université Libre de Bruxelles*. Already in 1908, Decroly mentioned Dewey as one of his primary sources of inspiration on the educational level.[37] Is it then surprising that Decroly was the only Belgian who translated Dewey, the French translation of *How We Think*:[38] *Comment Nous Pensons?*[39] Moreover, exemplary in the case of Decroly is the influence of Dewey's book *The School and Society*[40] in his daily educational praxis. Especially the motto "par la vie, pour la vie," the central idea in *The School and Society*, gives an impression of such an impact. Indeed, that pedagogical motto could be called Decroly's interpretation of this quotation from *The School and Society*: "Learning? Certainly, but living primarily, and learning through and in relation to this living."[41]

The question then arises of which ideas must be attributed to Dewey and which to Decroly. Did Decroly hinder the identification of the ideas of Dewey with the person of Dewey? Or, if you have read Decroly, why read Dewey? Decroly's case reveals the complexity of Dewey's work (his ideas and his language). The translation of *How*

We Think took about ten years of work, which was not just due to the war. In his introduction on *Comment Nous Pensons* as well as in several other chapters, Decroly pointed out many difficulties.[42] This complexity may have played an important role in the absence of identification of person and idea. Dewey himself complained that many of his followers—and especially among educators—either did not understand his work or garbled it just enough to use it for their own purposes.[43] It seems to us that he hit the nail on the head. Just as, for example, Darwinistic theories because of their ambiguity found shelter in the most divergent applications, so, too, Dewey's texts are open to several interpretations and are thus easy to implement in diverse social and ideological contexts. Contemporary commentators come to the same conclusion, like Richard Hofstadter who said: "Dewey was hard to read and interpret. He wrote a prose of terrible vagueness and plasticity."[44] And Martin Dworkin noticed: "Dewey wrote badly. His style was often opaque, his terminology ambiguous. (. . .) The problem of his language is inextricable from the problem of his philosophy, quite apart from the infelicities of his style." Many of Dewey's fans did not get the message, according to Dworkin: "It may be no compliment to professional educators," he said, "that they so easily understood Dewey while philosophers shook their heads."[45]

Paradoxically, the difficulties in complexity that have obstructed the dissemination may also have facilitated the diffusion of Deweyan ideas. It was exactly Dewey's vagueness and plasticity that contributed to the dissemination of his ideas. From this point of view, complexity has been a factor that helped the dissemination of Dewey's ideas. How do we reconstruct the process that took place? Besides the assumption that Dewey's ideas themselves didn't arise in a social, cultural, or intellectual vacuum (and thus have to be seen as an elaboration of other ideas), this means in the concrete, that his ideas were absorbed, fitted, assimilated, mixed, and identified with ideas of others. Moreover, they were filtered and implemented in new contexts apart from the American context—with their specific idiosyncrasies—without the conservation of the identification with Dewey. An example is the certain, but difficult to operationalize influence of *The School and Society* on Decroly's method. Indeed, Decroly used Dewey's ideas only to the extent that he could apply them in his own work.[46] Dewey was reduced to the "indigenous foreigner": he was an alien, a foreigner whose "language system" Decroly made his own as though he were a native.[47]

Ideas from Dewey were not just translated by Decroly but were also integrated in a set of "New Education ideas." Dewey became part of this "melting pot," his language fitting very well into Decroly's missionary, evangelistic, and propagandistic discourse of the first half of the twentieth century. In the *Introduction à l'histoire de l'éducation* by Arnould Clausse, both are even thrust forward as the figureheads of the *école nouvelle*, shackled to the slogan "école par la vie, école pour la vie."[48] The conclusion is that the authenticity problem between Dewey and Decroly actually is of little importance.[49] When we test this conclusion against textbooks by, for example, Frère Léon, it is confirmed. Although Dewey belonged in the bibliography to the classics that everybody had to have read—generally with the label of progressive scientist—[50] Dewey is no longer mentioned in the text: his name was not important; it was the message that counted.[51] The principle of *learning by doing* also underwent this transformation. Although it was generally mentioned in discussions of Dewey, it also took on its own life within the reformed pedagogical heritage.[52]

The Sliding of Dewey into the Pedagogical Canon

This transformation took on the character between the two world wars of sliding of the reform-pedagogical ideas ("child centered," pedocentric, *Vom Kinde Aus*, etc.)—from Dewey and others—into the "canon" of pedagogical ideas. The Catholic pedagogue De Hovre labeled the non-Catholic Dewey as someone whom educators could not ignore and also cited Dewey in the introduction of his study *'T Katholicisme: zijn paedagogen, zijn paedagogiek*.[53] In Daelemans and Van Hove, it sounds like this: "The principle of the activity has, since the work of Dewey . . . and others, become generally accepted in the new education along with other dynamic slogans."[54] We hear the same sounds also in the French-language publications: "one can say that it was truly in America with Dewey that this current of functional psychology and of scientific pedagogy in which we are immersed today in Belgium took shape in its concrete and positive form."[55] The integration of the New Education in pedagogical textbooks was, in other words, a success. From the reform-pedagogical position, too, the same sounds were made, but authors like D'Espallier expressed mostly astonishment that Dewey was able to settle in so quickly and so pervasively. In *Nieuwe banen in het onderwijs*, he states that it is "striking that the psychology of James and of that other pragmatist Dewey . . . penetrated so quickly." How profound this influence was is clear from the following example: "Without Dewey's psychology of thinking, without his instrumental logic, without his criteria for truth, the success of the material progress, the act, one can hardly under-stand the present school reforms from a psychological standpoint."[56]

These New Education elements were introduced, integrated, and adapted in the quest for a pedocentric Eldorado in order to support the thesis that pedagogy can save children.[57] A few examples, to give the floor to the opinion makers, can add force to this idea. So Marion Coulon stated: "For us, it is Dewey, the great American psychologist and educator who, on the threshold of modern times, first and most eloquently denounced the aberration of these traditional methods . . . with the ambition of thus preparing for a social life that would situate itself ever more under the sign of solidarity and collective effort."[58] Rombouts cites Dewey's figurative comparison of Dewey's work school with the Copernican revolution: "The child will henceforth be the sun around which the school institutions will turn, the middle point around which all of the education will be organized."[59] More crit-ical was Merecy, a state normal-school teacher in Lier, where he pointed to the lack of "*belles lettres* and artistic expression" and the excessive attention to "technical-industrial thinking."[60] To what extent could this be reconciled with the postulated pedagogical wonderland?

For all the great ideals, the history of the New Education teaches primarily that it, too, could not escape the demands of modernity and, even more, that its ideas had to be of service to this modernity[61]—in other words, that Dewey became a libation for modernity. Thus, pre-cisely as for Decroly, the concept of modernity occupied the first rank with Dewey, even to the point that Tom Popkewitz correctly calls him "an international spokesperson for the processes of modernization." The image of society that Dewey in *School and Society* took as the starting point was the new society, the modern and democratic society that was related to a particular reform Protestantism that he saw as the same as American democracy and which demanded far-reaching social changes. The most important change, which overshadowed and controlled all the others, was the industrial revolution.[62] The social effects had spread so quickly and so completely around us that Dewey deemed it inconceivable that education would only bear formal and superficial traces of it. In Belgium, it was primarily Omer Buyse who stressed Dewey's merits on the level of manual labor—*travaux manuels à caractère social*. This is understandable from the service that Dewey thus granted to modernity.[63] The fact that in the Belgian context the concept "society" could have had other coordinates and connections in constituting the modern was by the educational agents disseminating the "message" not taken into account.

Other pedagogical opinion makers also spoke in the same sense. Arnould Clausse agreed with R. Hubert, on the occasion of the Russian Revolution in 1918, that "the pedagogical revolution responded to an economic and technological necessity as well as to a political one."[64] Frère Léon warned the reader in his introduction to *Nieuwe wegen op: proeve van een nieuwe methodiek* that the object of his publication was "to help educators of good will forward, to bring those who wanted to advance themselves in contact with the new ideas."[65] Survival of the fittest—advancement under whatever form—was held before the curious reader as a reward, while Dewey himself preached general social progress.[66] That the principle "order in progress" remained an important theme is shown by the continuation: "We will apply these theories to public education in the spirit sketched above without faltering, without attacking anyone or anything but with the firm will to improve what is bad."[67] Van Coppenolle stressed this "order in progress" when he emphasized Dewey's American optimism: "I must frankly state that this American pedagogue was excessively optimistic and much too set against the old school and has committed many 'American exaggerations,' which has already become proverbial by its frequent occurrence."[68] Thus Dewey, as well as the reform-pedagogical discourse, became a libation for modernity.

That the insertion of the New Education into the Catholic camp did not necessarily constitute a problem was demonstrated by A. Daelemans, director-general of the *Nationaal Secretariaat van Katholiek Onderwijs* (the coordinating body of Catholic education), and Doctor W. Van Hove, teacher at the secondary Catholic normal school, in their *Algemene Onderwijsleer*: "Some philosophical assumptions, with which we do not necessarily have to agree, were at the foundation of the conceptions of these reformers, among others, pragmatism (Dewey)." In their *Algemene Onderwijsmethodiek*, they sought points of contact: "Finally, we find in our *philosophie* also a principle that points to the profound unity of body and soul, which together constitute the human *compositum*. Thought and action are together the factors of the human dynamic." However, he warns: "We have to avoid the excesses of an exaggerated manualism."[69] Regarding this exaggeration, Rombouts speaks of Dewey's one-sidedness.[70] All this leads to a sharp distinction, Dewey, so thought Daelemans and De Hovre, can be applied better to nature than to man, which also implies that the two can co-exist provided the terrain is thoroughly marked out and respected.[71] Thus far then the warning of Windey, Preter, and Pelgrims was kept in mind: "Does not one forget all too often that only the return to the centuries-old and forever young Christian doctrine of life can save the society? That only this doctrine can promote a healthy, balanced, and enduring renewal?" Moreover, these authors rely on the manual of De Keyser rather than their own reading of Dewey or visits to his schools.[72] Consequently, the Catholic religion took up the necessary task of guarding "order in progress" with regard to the social education of Dewey: "This system tries to develop itself outside of the Catholic religion. Therefore, it misses firm foundations and remains a pure experiment that will soon be thrust aside by a still more modern one."[73]

We are convinced that the study of transformation processes between higher and lower pedagogy (as noted in the example of De Keyser) offers insight into the size of the gap between theory and practice, but even much more in the manner in which the starting points were interpreted in function of the legitimation of the practice. There where we have linked the slowness of educational reform to its insertion into modernity, we also discussed the nature of the filter or transformation processes. Depaepe admits that the results of pedagogical didactic research "from the beginning of the twentieth century have been an *a posteriori* legitimation . . . of modernizing processes" and that researchers also felt the need to highlight the practical, future relevance.[74] In order to be able to be of service to modernity, sloganizing language was stripped from the underlying conceptual frameworks so that it became useful for everyone's purposes and could be integrated within its own structure. We encounter a striking example in Daelemans and Van Hove: "Although very many techniques of the education reformers are highly appreciated by us as important gains, we cannot concur with an obviously materialistic tendency in some New Education movements."[75] Arnould Clausse brings up for discussion the content of the great terms from the higher pedagogy. Thus, for example, it must be clear what the New Education means by "fulfillment in exploiting that which is fundamental in it," if not "one falsifies the truly human impact of the *Ecole Nouvelle* since it is placed at the service of an ideology that is the very negation of its objectives."[76] The question remains how this process precisely functioned. We give a number of indications of this below.

Penetration of Dewey

That the child-oriented New Education was able to place itself within a "meritocratic," neoliberal educational project of modernity stimulated the penetration but also detracted from the purity of the principles. Thus, the 1936 curriculum is generally considered as having been inspired by the New Education, but still the dilution of the principles "between the desire and the reality" is obvious. "The meager involvement of the teachers and of the representatives from the educational milieu in the decision-making process, the instability in the policy itself, and, not least, the high degree of personal tensions and rivalries among the policymakers" are important explanations for this gap.[77] In relation to the educational reform proposals around World War II, Olaf Moens, Frank Simon, and Jeffrey Tyssens came to the conclusion that the continuity of meritocratically inspired proposals constituted the theme both before and after World War II.[78] Indeed, the great revolution does not seem to have happened. That science legitimates also appears from the account of Dominique Grootaers: "the many references to the human sciences by the protagonists of comprehensive schooling actually mean a continuation of a concern of the representatives of the *Education nouvelle* from the previous two phases, namely the concern for a scientific basis for pedagogical practices. In the new approach, the same principles as from the psychology and the experimental pedagogy of the first half of the twentieth century regularly return: rationality and efficiency."[79]

Also with regard to the Flemish progressive pedagogical heritage after the 1960s, the diluted penetration can be linked to an insertion within the meritocratic, and in fact often neoliberal project of modernity, of which the educational policy amply witnessed since the 1970s.[80] Here, one cannot pass by the *Onderwijskrant*, since this progressive journal from its outset in 1977 has profiled itself as a critical forum for the progressive ideas of its own members. First of all, attention to progressive school models was replaced after awhile by interest in empirical research and psychological sophistication.With this, it is easier to legitimize one's own activities. For example, *Onderwijskrant* devoted a complete, albeit concise and critical article to Dewey. In fact, the article did not deal with Dewey but rather formed the background for the discussion of the legitimation arguments of constructivists on the basis of Dewey.[81] Second, the economic and cultural selection mechanisms in the alternative schools must be exposed: it is primarily the cultural elite of Flanders—those who can and will pay for a school of their own—who find shelter there while other groups of parents are represented to a much lesser degree. This group is not driven by a critical-emancipatory reflex to change society but rather by the hope that their children will acquire the critical skills needed to do well in the society of tomorrow. These are indications that Dewey and by extension the New Education in general were thereby so filtered that they served within the pedagogical melting pot to legitimate a meritocratic, neoliberal project, although New Education considered rhetorically paramount the emancipation of the child and with it also a social-democratic model of society.[82]

"From here you are on your own" (De Craene)

Instead of looking back on the foregone sections, we look forward and encourage further research. With recently acquired research material, the opportunities for the source research we proposed are central. The material consists of documentation that the Leuven Center for the History of Education received on the occasion of the retirement of Roland Vandenberghe, professor in educational innovation. It is of inestimable value for studying presence and influence. Even on the basis of a cursory examination, the documentation, which includes three studies of Dewey himself, offers a nuanced picture of the American. Further research should be able to determine whether or not Vandenberghe had still other works of Dewey and what picture of Dewey there emerges. Moreover, the study of intergenerational transfer is also a possibility, for Professor D'Espallier had given his books to Professor Vandenberghe. The passages marked by D'Espallier and Vandenberghe would then clarify the filter process. Carroi states in the introduction to *Expérience et éducation* that Dewey goes back to Rousseau and Froebel.[83] D'Espallier wrote in that same introduction "interesting" in the margin—where the text states "Agreeable and disagreeable lose their meaning; it's the motivation that counts"—and so unwittingly offers us an insight into his view of Dewey. The same applies in *Expérimentation en pédagogie* by Buyse with the indication "cfr. Dewey" for the sentence "To have ideas, to ask oneself questions, to devise hypotheses, to develop theories, to set up control experiments, to compare results, to draw conclusions—all this requires a new effort that engages other mental functions."[84] A wealth of opportunities is thus waiting with the object of acquiring knowledge about the filter processes that occur on the level of the pedagogical opinion makers.

In addition, there remains the question of the distance between the results of our investigation and the actual content of Dewey's publications. In this chapter, we have, with the concepts of "presence" and "influence" as the theme, only examined the reception and assimilation processes as regards the figure of John Dewey in the Belgian context. On the basis of exemplary source research, we came to the conclusion that Dewey is present as an "indigenous foreigner" in Belgium in the pedagogical "melting pot." Plucked from the "traveling library of ideas" and filtered through pedagogical opinion makers, the thoughts of this American philosopher and educator were used—and misused—as a sacrifice on the altar of the neoliberal modernity whereby unmistakably a gulf developed between the original, innovation-oriented ideas of Dewey and their ready-made application in the renewal-shy environment of education. So what actually remained of all that rhetoric as soon as the practitioners, who at most had to learn fragments by heart to earn their diploma, stood in the classroom? Probably very little at all. This applies *a fortiori* for the contemporary situation, for Dewey stands, like all the other gods of the pedagogical pantheon, hardly at all on the education menu of the teacher training colleges. In the present successful period of free-market thought, the neoliberal educational project seems to have become so self-evident that the average educational worker did not want to have anything to do with this theory.

Nevertheless, this does not exclude the desirability of research in the direction suggested in this chapter. Quite the contrary, for only the critical unraveling of the high-sounding words that pedagogues of every sort use to legitimate their agendas, hidden or not, can help in the search for a more "realistic" expectation with regard to education. And in this historiography and its nuisance value can be a powerful partner.

Notes

1. For example: Frans De Hovre, *Paedagogische wijsbegeerte: een studie in de moderne levensbeschouwingen en opvoedingstheorieën*, Malmberg's Paedagogische Bibliotheek, Vol. 14 ('s-Hertogenbosch and Antwerp: L.C.G. Malmberg, 1924), 94; Frans De Hovre and L. Breckx, S.d. *Les maîtres de la pédagogie contemporaine* (Bruges: Charles Beyaert; Paris: Casterman), 35; Frans De Hovre, *'T Katholicisme. Zijn paedagogen. Zijn Paedagogiek*. Malmberg's Paedagogische Bibliotheek, Vol. 17 ('s-Hertogenbosch: L.C.G. Malmberg, 1925), XXIV.
2. With regard to the anglophilia in De Hovre, see Marc Depaepe, "Katholische und nationalsozialistische Pädagogik in Belgien, 1919–1955. Ihre ambivalente Beziehung im Spiegel der 'Vlaamsch Opvoedkundig Tijdschrift.' " *Zeitschrift für Pädagogik* 44, no. 4 (1998): 504–522. In relation to the non-Catholicism in Dewey, see Jay Martin, *The Education of John Dewey: A Biography* (New York: Columbia University Press, 2002), 19–25, 147–149, 163–164.

3. Marc Depaepe, *Gesplitst of gespleten? De kloof tussen wetenschappelijke en praktische kennis in opvoeding en onderwijs* (Louvain and Leusden: Acco, 2002), 37.

4. Marc Depaepe, *De pedagogisering achterna. Aanzet tot een genealogie van de pedagogische mentaliteit in de voorbije 250 jaar* (Louvain and Amersfoort: Acco, 1998); Marc Depaepe, Kristof Dams, Maurits De Vroede, Betty Eggermont, Hilde Lauwers, Frank Simon, Roland Vandenberghe and Jef Verhoeven, *Order in Progress. Everyday Educational Practice in Primary Schools: Belgium, 1880–1970*, Studia Paedagogica, n.s. 29 (Louvain: Leuven University Press, 2000), 13.

5. Marc Depaepe, "De spanning van verborgen agenda's. Ontwikkelingen inzake het onderwijs in de historische pedagogiek," *Comenius. Wetenschappelijk Forum voor Opvoeding, Onderwijs en Cultuur* 17, no. 3 (1997): 200–215.

6. Marc Depaepe, "The practical and professional relevance of educational research and pedagogical knowledge from the perspective of history: reflections on the Belgian case in its international background," *European Educational Research Journal* 1, no. 3 (2002): 360–379. Electronic text available at http://www.wwwords.co.uk/eerj/.

7. Jo Ann Boydston, ed., *Guide to the Works of John Dewey* (Carbondale and Edwardsville: Southern Illinois University Press; London and Amsterdam: Feffer and Simons Inc, 1972).

8. Susan Semel, F. "Women progressive leaders in the twentieth century: From Caroline Pratt and Helen Parkhurst to Lillian Weber and Deborah Meier," Paper presented at the 26th annual meeting of the ISCHE at University of Geneva, Geneva, July 14–17, 2004.

9. Marc Depaepe, Frank Simon, and G. Verbeek, "Vom französischer Dominanz zur kulturellen Autonomie. Sprachproblematik und Unterricht im flämischen Teil Belgiens (1830–1990)," Für Prof. Dr. Maurits De Vroede, *Zeitschrift für Pädagogik* 40, no. 1 (1994): 97–112.

10. Sol Cohen, *Challenging Orthodoxies. Toward a New Cultural History of Education* (New York: Peter Lang, 1999), 89.

11. See, e.g.: Jaime Caiceo Escudero, "Algunos antecedentes sobre la persencia de la escuela neuva en Chile durante el siglo XX," Paper presented at the 26th annual meeting of the ISCHE at University of Geneva, Geneva, July 14–17, 2004; Angelo Van Gorp, "Gedragswetenschap in de steigers. Het psycho-pedagogisch vertoog van Ovide Decroly ontmythologiseerd? (1871–1932)" (Dissertation, Katholieke Universiteit Leuven, Louvain, 2003), 1–32, 263–316.

12. Frater Sigebertus Rombouts [Godfried Frans Rombouts], *Historiese pedagogiek. Grote lijnen der geschiedenis van het opvoedkundig denken en doen in doorlopend verband met de kultuurontwikkeling*, 4 vols. (Tilburg: RK Jongensweeshuis; Amsterdam: RK Boekcentrale; Antwerp: NV Veritas, 1928); Frater Sigebertus Rombouts [Godfried Frans Rombouts] *Leerboek der historiese pedagogiek* (Tilburg: RK Jongensweeshuis; Amsterdam: RK Boekcentrale; Antwerp: NV Veritas, 1928).

13. Depaepe, *Gesplitst of gespleten?* 16.

14. Depaepe, *Gesplitst of gespleten?*; Tom De Coster. "Gefilterd verleden: over de afstand van wetenschappelijke en schoolse kennis in het geschiedenisonderricht in de lagere scholen van het interbellum" (Lic. Thesis, Katholieke Universiteit Leuven, Louvain, 2002); Jürgen Schriewer, "Internationalisierung der Bildungsforschung und nationale Reflexionstraditionen," in *Science(s) de l'éducation 19ᵉ–20ᵉ siècles. Entre champs professionnels et champs disciplinaires/ Erziehungswissenschaf & (en) 19.–20. Jahrhundert. Zwischen Profession und Disziplin*, ed. R. Hofstetter and B. Schneuwly, 507–527. Exploration: Education: histoire et pensée (Bern e. a.: Peter Lang, 2002).

15. Frans De Hovre, *Paedagogische denkers van onzen tijd. Bloemlezing* (Antwerp, Brussels, Ghent and Louvain: N.V. Standaard-Boekhandel, 1935), 13–15, 34.

16. See, e.g.: Jan Baert "De aandacht voor John Dewey in het Belgisch pedagogisch milieu (1920–1940): een eerste verkenning" (Lic. Thesis, Katholieke Universiteit Leuven, Louvain, 1982), 194–203.

17. Depaepe et al. *Order in Progress*, 44.

18. We only know of a licentiate thesis: See Liesje Raemdonck, "De belgische academicus tijdens het interbellum. Een veldanalyse van de professoren van de faculteit Letteren en Wijsbegeerte van de vrije universiteiten" (Lic. Thesis, Vrije Universiteit Brussel, Brussels, 1996–1997).

19. Frederik Lilge, "Dewey, John," in *Standaard Encyclopedie voor Opvoeding en Onderwijs*, Vol. 1, ed. Alfred De Block et al. (Antwerp: Standaard; Hoorn: Kinheim, 1974), 392–393; P. A. Hoogwerf, "Dewey, John," in *Paedagogische Encyclopaedie*, Vol. 1, ed. Jozef E. Verheyen and Rommert Casimir, 350–355 (Antwerp: De Sikkel, 1939), 354; Victor D'Espallier, "Dewey, John," in *De Katholieke Encyclopaedie van Opvoeding en Onderwijs*, Vol. 1, ed. Victor D'Espallier et al., 458–461 ('s Gravenhage: Pax; Antwerp: 't Groeit, 1951), 461; Cyriel De Keyser, *Inleiding in de geschiedenis van het Westerse vormingswezen* (Antwerp: Plantyn, 1958), 440.

20. John Dewey, *School en maatschappij*, trans. from the English by T. J. de Boer (Bekende paedagogen, Groningen and The Hague: J.B. Wolters' Uitgevers-Maatschappij N.V, 1929).

21. John Dewey, *Ervaring en opvoeding*, trans. from the English by Gert Biesta and Siebren Miedema (Houten and Diegem: Bohn Stafleu Van Loghum, 1999).

22. Coming then in the picture, inter alia, are: G. Wielenga, "Dewey," in *Grote denkers over opvoeding*, ed. I. Van Der Velde (Groningen: E. P. Noordhoff e.a, 1964), 309–336; N. F. Noordam, "John Dewey," in Onderwijskundigen van de twintigste eeuw, ed. Q. L. van der Meer and H. Bergman (Groningen: Wolters-Noordhoff, 1975), 15–22; Frater Sigebertus Rombouts [Godfried Frans Rombouts], *Historische pedagogiek*, 7th ed. (Tilburg: Uitgeverij Zwijsen, 1958); Louis M. A. N. Van Schalkwijk, *De sociale paedagogiek van John Dewey en haar filosofiese grondslag. Academisch proefschrift door Louis Marthinus Albertus Nicolas van Schalkwijk* (Amsterdam: 't Portiekje, 1920); Joop W. A. Berding, *De participatiepedagogiek van John Dewey: opvoeding, ervaring en curriculum* (Leiden: DWSO Press, 1999).

23. For a survey of French translations till 1967, see: Gérard Deledalle, *L'idée d'expérience dans la philosophie de John Dewey*, Publications de l'Université de Tunis, Faculté des lettres et sciences humaines 6, no. 1 (Paris: Presses Universitaires de France, 1967), 538–539.

24. Boydston, *Guide to the Works of John Dewey*.

25. John Dewey in the index of names of: Maurits De Vroede et al. *Bijdragen tot de geschiedenis van het pedagogisch leven in België in de 19de en 20ste eeuw. Deel I: De periodieken 1817–1878; Deel II: De periodieken: 1878–1895; Deel III: De periodieken 1896–1914; Deel IV: De periodieken 1914–1940* (Louvain: Universitaire Pers, 1974–1987).

26. Maurits De Vroede et al., *Bijdragen tot de geschiedenis van het pedagogisch leven in België in de 19de en 20ste eeuw. Deel IV: De periodieken 1914–1940 Eerste stuk* (Louvain: Universitaire Pers, 1987), 476.

27. This conclusion is also raised in: Jan Baert, "De aandacht voor John Dewey in het Belgisch pedagogisch milieu" (1920–1940), 194–203.

28. Frederik Lilge, "Dewey, John," 393; P. A. Hoogwerf, "Dewey, John," 354; Victor D'Espallier, "Dewey, John," 461; Frans De Hovre, *Paedagogische denkers van onzen tijd. Bloemlezing*, 34; Marcel A. Nauwelaerts, *Opvoeding en onderwijs in het verleden: teksten en documenten. Deel 2. 2d. ed.* (Louvain: Acco, 1977), 136–140; Cyriel De Keyser, *Inleiding in de geschiedenis van het Westerse vormingswezen*, 440.

29. Ministère des Sciences et des Arts/Ministerie van Kunsten en Wetenschappen, *Rapport triennal sur la situation de l'instruction primaire en Belgique présenté aux Chambres Législatives le 9 mars 1928 par M. M. Vauthier/Verslag over den staat van het lager onderwijs in België op 9 maart 1928 aan de Wetgevende Kamers aangeboden door den Heer M. Vauthier* (Brussels: Ministère des Sciences et des Arts/Ministerie van Kunsten en Wetenschappen, 1928), 28.

30. Marcel A. Nauwelaerts, *Opvoeding en onderwijs in het verleden*, 136–140.

31. For example, about the material and symbolical meaning of the pencil, see: Henry Petroski, *The Pencil: A History of Design and Circumstance* (New York: Alfred A. Knopf, 1993).

32. P. A. Hoogwerf, "Dewey, John," 351–353.

33. Raymond Buyse, "Voyage d'Etude aux USA" (Diary, Archives of Professors, University Archives K.U. Leuven, Central Library, Louvain, 1922).

34. Paraphrased in: Ovide Decroly and Raymond Buyse, "Le rêve entrevu. Une journée à 'Park School'" (Buffalo, United States). *Pour l'Ere Nouvelle* 1, no. 4 (October 1922): 70.

35. Ovide Decroly and Raymond Buyse, *Les applications américaines de la psychologie à l'organisation humaine et à l'éducation* (Documents Pédotechniques, vol. 2.2. Brussels: Lamertin, 1923); About the interpretation of Dewey's reaction: see Jay Martin, *The Education of John Dewey: A Biography* (New York: Columbia University Press, 2002); Robert B. Westbrook, *John Dewey and American Democracy* (Ithaca, NY: Cornell University Press, 1991).

36. Guy Montrose Whipple, ed., *The Twenty-Eighth Yearbook of the National Society for the Study of Education. Preschool and Parental Education* (Bloomington, IL: Public School Publishing Company, 1929), 17; Angelo Van Gorp, Gedragswetenschap in de steigers, 332.

37. Ovide Decroly, "Le programme d'une école dans la vie," 362, *L'Ecole Nationale* 7, no. 11 (1908): 323–325; 7, no. 12 (1908): 360–362.

38. John Dewey, *How We Think* (Boston, New York and Chicago: D.C. Heath and Co., 1910).

39. John Dewey, *Comment nous pensons*, trans. from the English by Ovide Decroly (Bibliothèque de Philosophie scientifique. Paris: E. Flammarion, 1925).

40. John Dewey, *The School and Society* (Chicago: The University of Chicago Press, 1899).

41. John Dewey [*The Child and the Curriculum and*] *The School and Society. Introduction by Leonard Carmichael* (Chicago and London: The University of Chicago Press, 1971); John Dewey, *The School and Society*, in *Dewey on Education: Selections*, 7th ed, ed. Martin S. Dworkin, Classics in Education, Vol. 3 (New York: Teachers College Press, 1971), 34–88.

42. Dewey, *Comment nous pensons*, 284; Dewey, *How We Think*.

43. Dewey, *The School and Society*, 14.

44. Richard Hofstadter, *Anti-Intellectualism in American Life* (New York: Alfred A. Knopf, 1979), 361.

45. John Dewey [*The Child and the Curriculum and*] *The School and Society*, 13–14.

46. William J. Reese, "The origins of progressive education," *History of Education Quarterly* 41, no. 1 (2001): 10.

47. Sol Cohen, *Challenging Orthodoxies: Toward a New Cultural History of Education* (New York: Peter Lang, 1999), 89.

48. Arnould Clausse, *Introduction à l'histoire de l'éducation* (Brussels: Maison d'édition A. De Boeck, 1951), 18.

49. Regarding the exchangeability of the ideas of Decroly and Dewey, see: Angelo Van Gorp, "Gedragswetenschap in de steigers," 330–333: Tobie Jonckheere, *Fragments d'une histoire de l'éducation: quelques noms et quelques faits* (Brussels: De Boeck, 1958), 227, 247, 249.

50. Cyriel De Keyser, *Inleiding in de geschiedenis van het Westerse vormingswezen*, 440; Victor D'Espallier, "Dewey, John," 458; Frederik Lilge, "Dewey, John," 392.

51. For example: Frère Léon [Jules Cyriel Lust], *Leçons de psychologie appliquée à l'éducation* (Paris: Desclée and De Brouwer, 1943).

52. For example: Victor D'Espallier et al., *Nieuwe banen in het onderwijs*, 2nd. ed. Vol. 1 (Antwerp, Brussels, Ghent and Louvain: Standaard Boekhandel, 1937), 121; Frère Anselme [Joseph D'Haese], *Pour enseigner mieux—méthodologie*, 167, 192, 295.

53. For example: Frans De Hovre, *Paedagogische wijsbegeerte*, 94.

54. Alfred Daelemans and W. Van Hove, *Algemene onderwijsmethodiek*, (Hoboken: Uitgeverij Plantijn NV, 1954), 65.

55. Marion Coulon, *Jeunesse à la dérive*, Vol. 5 and 6, *Pour un nouveau statut général de l'enseignement et du personnel enseignant* (Mons: Silène, 1948), 49.

56. Victor D'Espallier et al., *Nieuwe banen in het onderwijs*, 8.

57. Depaepe, *Order in Progress*, 11–19; Marc Depaepe, *De pedagogisering achterna*. Depaepe, 13–42; encyclopedias have stressed the contrast between

"old" and "new" education in the discourse of the New Education: Marcel A. Nauwelaerts, "Nieuwe-schoolbeweging," in *Standaard Encyclopedie voor Opvoeding en Onderwijs*, Vol. 4, ed. Alfred De Block et al. (Antwerp: Standaard; Hoorn: Kinheim, 1974), 19–20; Rombouts, Frater Sigebertus [Godfried Frans Rombouts]. 1951. Nieuwe-Schoolbeweging. In *De Katholieke Encyclopaedie van Opvoeding en Onderwijs*, Vol. 3, ed. Victor D'Espallier et al. ('s Gravenhage: Pax; Antwerp: 't Groeit, 1951), 236–238.

58. Marion Coulon, *Jeunesse à la dérive*, 90.

59. Rombouts, Frater Sigebertus [Godfried Frans Rombouts], *Historiese paedagogiek. Grote lijnen der geschiedenis van het opvoedkundig denken en doen in doorlopend verband met de kultuurontwikkeling*, Vol. 3 (Tilburg: RK Jongensweeshuis; Amsterdam: RK Boekcentrale; Antwerp: NV Veritas, 1928), 249, 265.

60. Renaat Mercy, *Historische pedagogiek. Schets van ideeën en werkelijkheden: Hellas tot heden* (Antwerp: De Sikkel, 1966), 144.

61. Depaepe, *Gesplitst of gespleten?* 40.

62. Dewey *[The Child and the Curriculum and] The School and Society*, 9.

63. With regard to Buyse: Raymond Buyse, *Méthodes américaines d'éducation générale et technique*, 3rd. ed. rev. and enlarged. Paris: H. Dunod & E. Pinat, 1913), 141. With regard to the section: Angelo Van Gorp, "Gedragswetenschap in de steigers," 317–333.

64. Arnould Clausse, *La relativité éducationnelle. Esquisse d'une histoire et d'une philosophie de l'école* (Paris: Fernand Nathan; Brussels: Editions Labor, 1975), 45; see also: Arnould Clausse, *Introduction à l'histoire de l'éducation*, 132, 143.

65. Broeder Leon [Jules Cyriel Lust], *Nieuwe wegen op. Proeve van een nieuwe methodiek*, Vol. 1, *Algemene methodiek* (Brussel: Broeders Maristen, 1937), 6.

66. See e.g.: Clausse, La relativité, 175–176.

67. Broeder Leon [Jules Cyriel Lust], *Nieuwe wegen op*, 6.

68. Valère Van Coppenolle, *De activiteit op school. Bondig historisch overzicht.* Opvoedkundige Brochurenreeks van de Studiekringen van het Christen Onderwijzersverbond, Vol. 12 (Torhout: Pyck, 1939), 52; With regard to exaggerations, see: R. Windey, K. De Preter, and M. Pelgrims, *Geschiedenis van opvoeding en vorming met bloemlezing*, 4th ed. (Antwerp: Plantyn, 1965), 178–179.

69. Alfred Daelemans and W. Van Hove, *Algemene Onderwijsleer* (Hoboken: Uitgeverij Plantijn NV, 1954), 41.

70. Frater Sigebertus Rombouts [Godfried Frans Rombouts], *Historiese paedagogiek. Grote lijnen der geschiedenis van het opvoedkundig denken en doen in doorlopend verband met de kultuurontwikkeling*, Vol. 4 (Tilburg: RK Jongensweeshuis; Amsterdam: RK Boekcentrale; Antwerp: NV Veritas, 1928), 204.

71. Daelemans and Van Hove, *Algemene onderwijsmethodiek*, 65–66, 74.

72. Windey et al., *Geschiedenis van opvoeding en vorming met bloemlezing*, 178–179.

73. Désiré Pissens, *Geschiedenis van onderwijs en opvoeding voor normalisten*, 2nd ed. (Brussels: Vrije Normaalscholen, 1925), 132–133.

74. Depaepe, *Order in Progress*, 11–17, 249; Depaepe, *Gesplitst of gespleten?* 9–16.

75. Daelemans and Van Hove, *Algemene onderwijsleer*, 69.

76. Clausse, *La relativité educationnelle*, 42–43.

77. Marc Depaepe, Maurits De Vroede, and Frank Simon, "The 1936 curriculum reform in Belgian primary education," *Journal of Education Policy* 6, no. 4 (1991): 371.

78. For example, the "école unique" idea of general primary education, a forerunner of comprehensive secondary education; see: Olaf Moens, Frank Simon, and Jeffrey Tyssens, "De dag van de opvoeders is nu op komst: onderwijshervormingsvoorstellen rond de Tweede Wereldoorlog," in *De Tweede Wereldoorlog als factor in de onderwijsgeschiedenis/La Seconde Guerre mondiale, une étape dans l'histoire de l'enseignement*, ed. Marc Depaepe and Dirk Martin (Brussels: Navorsings- en Studiecentrum voor de Geschiedenis van de Tweede Wereldoorlog/Centre de Recherches et d'Etudes Historiques de la Seconde Guerre Mondiale, 1997), 57–58.

79. Dominique Grootaers, "Belgische schoolhervormingen in het licht van de 'Education nouvelle' (1870–1970)," in *Reformpedagogiek in België en Nederland*, ed. Nelleke Bakker, Pieter Boekholt, Hans Van Crombrugge, Marc Depaepe, and Frank Simon, 9–33, Jaarboek voor de geschiedenis van opvoeding en onderwijs (Assen: Van Gorcum, 2001), 26–27.

80. For example with the "redefining" of the idea of comprehensive education in secondary schools; see: Bregt Henkens, "The rise and decline of comprehensive education: Key factors in the history of reformed secondary education in Belgium," *Paedagogica Historica* 40, no. 1 and 2 (2004): 193–209.

81. Raf Feys, "John Dewey: boegbeeld van progressive education," *Onderwijskrant* 118, November: 18–22.

82. Lieven Viaene, "Kiezen voor een niet traditionele basisschool. Een exploratieve studie naar het recruteringsveld van de alternatieve basisscholen in Vlaanderen en naar de keuzemotieven van de ouders" (Lic. thesis, Vrije Universiteit Brussel, Brussels, 1992); Dirk Dieltjens, "Tussen utopie, rebellie en seculaire religie. Een biografisch onderzoek naar belief systems bij emancipatorische onderwijsvernieuwers in de Werkgroep Basisschool van Aktiegroep Kritisch Onderwijs" (Lic. thesis, Vrije Universiteit Brussel, Brussels, 1990–1991); Tom De Coster, "Wat is er in Vlaanderen met het progressief pedagogische erfgoed gebeurd? Inspiratiebronnen en hun filterproces in het kader van receptie- en implementatiegeschiedenis van 'emancipatorische' opvoedingsmodellen in de neoliberale context na de jaren 1960," Paper presented at the 7th annual meeting of the Association des Cercles Francophones d'Histoire et d'Archéologie de Belgique and the 44th annual meeting of the Fédération des Cercles d'Archéologie et d'Histoire de Belgique at University of Louvain-la-Neuve, Louvain-la-Neuve, August 26–27, 2004.

83. John Dewey, *Expérience et éducation*, trans. from the English by M. A. Carroi, Educateurs d'hier et d'aujourd'hui (Paris: Editions Bourrelier and Cie, 1947), 13, 17.

84. Raymond Buyse, *L'expérimentation en pédagogie* (Brussels: Maurice Lamertin, 1935), 122.

5

Dewey on Lima or the Social Prosthesis in the Construction of the New Education Discourse in Portugal (1925–1936)

Jorge Ramos do Ó

Introduction

This chapter discusses the use of John Dewey in the educational discourse of Adolfo Lima (1874–1943). Lima was a fundamental pedagogue in the reception, structuring, and re-launching of the so-called New Education in Portugal. His action has been highlighted by the historiography of education,[1] alongside figures such as Faria de Vasconcelos (1880–1914), Álvaro Viana de Lemos (1881–1972), and António Sérgio (1883–1969). While these three can take most of the scientific, philosophical, and organizational credit for the movement,[2] it was Adolfo Lima who articulated the reforming ideas to transform the school landscape. He was a teacher trainer, secondary school-teacher, and pedagogical director of private and public institutions, where he implemented innovative educational experiments at the start of the twentieth century in Portugal. Furthermore, he founded and ran a pedagogical reflection journal, published a methodology and didactics manual for teachers, and, in addition to a host of essays about education, even published a pedagogical encyclopaedia.

Adolfo Lima did much more than merely disseminate the work of John Dewey in Portugal from the mid-1920s onward. Certainly without Lima, Dewey would have remained practically unknown in Portugal for at least the first half of the twentieth century.[3] But the essence of Lima's work was not this. I argue in this chapter that Dewey functioned as a reference point that enabled Lima both to structure his entire educational discourse and to put forward proposals to transform Portuguese schools, which were immediately deemed credible by the different school authorities.

John Dewey as a Germ and Crystal of Adolfo Lima's Pedagogical Discourse

In the formation of Lima's differentiated thinking, with regard to his peers in Portugal, Dewey played a decisive role, transforming into the *conceptual personage* of Adolfo Lima. His educational ideas only effectively made global sense when associated with the conceptual landscape created by the American philosopher. The expression "conceptual personage" is used here in the sense that Deleuze and Guattari[4] attributed to a skill of thinking that operates in the fields of immanence of the author and intervenes in the creation of his concepts. Dewey was therefore the "heteronym" and enunciation agent, always complying with this role of manifesting the "territories" and the "reterritorializations" of the Portuguese educator.

John Dewey's name must also be associated with the emergence in Portugal of truly innovative organizational solutions that, through the educational practices implemented by Lima, had a big impact on the public educational system from the 1930s onward. The historical construction of national discourses and practices should be understood, as pointed out by Popkewitz,[5] in the light of a hybridization and a symbiosis, in which the foreign hero becomes part of the indigenous culture. Dewey did indeed function for Lima as the indigenous foreigner able to lend his discourse a strong operative force: Lima claimed educational innovation based on a scientific framework able to separate the boundaries of reason and non-reason and to construct social progress. He introduced, and in a certain way became the unique owner, of the prosthesis of political subjectivation in the pedagogical renovation discourse in Portugal. He elaborated in detail on the topic of the school as a *miniature society* to project the practical scenario in which the ideal independent-responsible-free child would be formed: By only exercising a specific kind of political power, he would learn to interiorize a set of rules of conduct that will enable both to promote his spontaneous adaptation to school life and his ability to self-govern.

Always backed up by Dewey, Lima was also one of those pedagogues who argued that the traditional school model would only be superseded when all the methods and techniques were adapted to the reality of each child, that is, the child's innate laws and particular interests. But it was gradually understood that the core of educational reforms involved finding an organizational scenario in which the child's idiosyncrasy could be transformed into a calculable and governable force. The education through measurement advocated by psycho-pedagogy was recognized as extremely important, but not entirely sufficient to bring about the sought-after change to the paradigm. Lima wanted to be seen as the author of an original proposal when he defended the theory that the traditional school model can only be supplanted by another that placed the *social* over the *psychological*. Indeed, he believed that the school should not merely understand and act on the psychological soul. The training of the pupil required political subjectivation, that is, the reproduction within the school of political institutions. Therefore, the bulk of his efforts focused on conceiving the school as a project of power in which the principles of a pedagogy exclusively based on psychology and individual morality were supplanted by the dynamics of work and civism, always aimed at social integration. Lima viewed school as an organization completely indistinct from any other institution developed by the modern state. Its utopia would derive from the pragmatism of Dewey, as it seemed to him necessary to invent an organizational formula that merged the interior and exterior of the child via the child's social duty and personal interest. What is fascinating is that in Portugal this dynamic of self-government of the child interested the authorities, the liberals and the conservatives alike.

Criticize the Model of the Faculties of the Soul

Lima included a sentence in the first volume of *Metodologia* written by John Dewey. This helps us to begin to understand, at a glance, how the pedagogical discourse of the Portuguese author attempted to achieve global intelligibility using the legacy of the American philosopher as his starting point. The sentence reads: "repelling the doctrine of the *faculties of the soul* and adopting 'the functionalist conception, as corresponding more closely to the complete evolution of the child,' manifestation of the existence of a psychic state or conscience is in the *interest* shown by the child with regard to the quality and quantity of knowledge or education that he would like to have."[6]

This citation legitimizes a notion of frontier in the educational field, borne out of a new cultural image of childhood. Only the psychic interiority of the pupil was mentioned, affirming that all the work to be carried out by the school institution depends on the willingness or desire of the child. Although referring to the processes of integration into social life, Lima wrote with the certainty of one who was redesigning an educational metanarrative.

On several occasions, Lima appropriated from Dewey the expression "Copernican revolution" to identify the kind of change that should occur in the educational field. While Copernicus shook the foundations of astronomy by showing that our planet was not at the center of the universe, one has to recognize in the author of *The Child and the Curriculum* the proposal for a deep-rooted and radical inversion of school values.

The crux of the change as defined by Dewey is: Where previously one would read teacher, one should now read pupil. "Today it is acknowledged as a mistake," continued Adolfo Lima, "to consider the master as the centre, around which all school life revolves and is subordinated"; the new centrality around which all school activity should gravitate has to be the child. "Blind and absolute" subordination to the teacher was no longer valid and it is the job of the teacher to adapt to the demands of the nature of the child, so that, "in going through the metamorphoses necessary for the expansion of his later life" the child develops completely "physiologically, aesthetically, mentally and socially" as "a personality conscious of his duties and rights."[7]

This new centrality would have a big impact on the various foundations of the school system. The new paradigm obviously implied a frontal refusal of both the current conceptions of childhood and the traditional school model, which was completely operative at the time. In general, children continued to be looked upon as "things" without their own character, and hence adults maintained their complete omnipotence, selecting for the pupils "the studies and games, books and toys, occupations, leisure activities and professions." They continued to "command, govern, order, admonish, censure and punish." "School functioning has been subject to, and still is subject to," explained Adolfo Lima in his *Pedagogia Sociológica*, "the exclusive conveniences and whims of the teachers." As such, the timetables, organization of the classes, the discipline and the distribution of the service as a whole were still "subordinated to the pure autocratic arbitration of the commanding adult." In his opinion the disciplinary-repressive empire of the "prison-school," the "barracks-school," or the "convent-school" with its "ancient school customs" had not yet been supplanted.[8]

In an obvious contradiction "to the aspirations of modern social life," the school institution therefore continued to "pull apart and shred to pieces for ever all aspects of an individuality," making the child "an automaton, a slave, a half-wit, a sheep able to be tamed by the shepherd."[9] For Lima, all this "passivity" was fed by the "intellectualism"— this "abusive and sterile tendency towards an education through formalist education, without an objective goal."[10] The era of the citizen and the "social man" had therefore not yet reached the educational territory.[11]

But in the preceding decades and owing to the "development of the teaching of Psychology and the evolution of societies," a worldview of childhood and a rationality of the educational relation began to be structured that could dethrone centuries-old conceptions and practices.[12] And here Lima's criticism became less general and was directed at the prior generation of pedagogues. Obviously, it was the process of replacing legitimacies, aimed at bringing about the downfall of the theories of Buisson, Compayré, and Marion, names that had been associated from the last quarter of the nineteenth century, in the establishment of Education Sciences as an autonomous disciplinary field of knowledge.

These authors, among several others, articulated a pedagogical discourse based on Philosophy that, although claiming a *psy* framework, kept intact the idea of the Cartesian–Kantian subject, that is, absolutely fixed to the universes of reason around the so-called *faculties of the soul*.[13] For them, it was child psychology that would respond to the need to ascertain the three faculties of "sensibility," "will" and "intelligence" because this science acknowledged the diversity of individual characters. Henri Marion provided an appropriate definition of the discipline: "psychology means science of the soul: the field of psychology changes according to the way one understands the soul and according to whether one accepts that there can be a scientific knowledge of it."[14] Compayré explained: "Pedagogical action in the field of the faculties of the soul should come as close as possible to the order of nature. In this way an evolution is favoured that leads from the concrete to the abstract, from instinctive life to reflective life; in this way the faculties of the soul gain their own activity, a dynamism and an energy that will allow them to increasingly develop by themselves throughout life; therefore, school education can be successful in all ages through personal education, self-education."[15]

A New Pedagogical Legitimacy Based on Psychic Economy and on the Interest of the Child

Lima returned to his main source to begin to explain arguments of a scientific nature able to demonstrate the distinct *psychological body* of each child. At the helm of the American psychologists was again Dewey, who internationally was the face against the formal culture and the doctrine of the faculties. In the words of Lima, Dewey's modern psychological science, based on experimentation, was able to guarantee that thought, reason, or observation could not continue to be viewed as "uniform powers, existing in themselves, but rather as multiple paths through which a need could be developed in an action. They would have to be understood as forms of adaptation that varied according to the circumstances of each child and the respective surrounding environment. Lima clarified the sense of change of the educational paradigm:

> Thinking, says Dewey, "is the power of coordinating the particular suggestions that particular things arouse into a group." The representatives of this functionalist conception are, however, far from wanting to lower education from the spirit to an apprenticeship. They believe just the opposite; that it is by constantly connecting mental work to needs, to tasks defined so as to arouse the interest of the child, that these habits of thinking shall be formed. This will develop in the child, little by little, the techniques that become more complicated as he ages, which will enable the child to solve problems put before him.[16]

John Dewey's position in today's philosophical debate is decisive to scientifically validate the point of view that educational action is only effective if it is presented as a work of subjectivation borne out of the subconscious fantasy of the child and then returns to the child in the organized form of a *habitus* of his thinking. It is as if the outside/inside dichotomy disappears in this dynamic of an action on the mental action, and if the self-consciousness would lead to self-creation.

According to Lima, contemporary philosophy led to "two orientations." One, belonging to the intellectuals, who, "excited by the progresses of science," were convinced that it was "exclusively through science that the human being could obtain *certainty*." The other, belonging to the philosophers that were not "satisfied with science, demanding more from it than it could give." This second group did not belittle science itself, but opposed "scientific exclusivity, the pure positivists." Among them, of particular note was "American pragmatism" and within this "the instrumentalisation of J. Dewey." If one continued to admit that science served to "ensure effective power over nature," this second current postulated that it was exactly "at the irrational heart of our being, often even in the subconscious, that we must search for who we are and what nature is." Lima continued to criticize the empire of rationalism, accepting the possibility of a theory that one reached the public sphere starting from the private space: "the aspirations of our heart, our most obscure instincts, know far more than the clarified truths of reason." True knowledge should be sought in other places, "in our moral ideas, in our sentimental intuitions." It was important that the "cerebral physiology" came into play in the experimental, and only Psychology, "as an applied science and determinist of the soul," could reasonably explain states of conscience as "sensations, desires, commotions, knowledge, reasoning, decisions or violations."[17]

Therefore, it was based on the "applied" knowledge of the child that a starting point should be established for the new education. Lima viewed this question of observation as a permanent task that was never entirely settled. The pedagogue could not for a single moment neglect to follow "the various processes of natural development of the child, to be able to surprise the most beneficial psychological moments of action for determined kinds of teaching and education." This task was extremely delicate because it involved acting on spontaneity and establishing the criteria of internalization of social practices, far from any external, rational, and conscious point of support. He explained his argument in a good deal of detail, dealing with the problem of childhood attention. This aspect provided him with a good example to justify his theories. It seemed to him undeniable that the child was, "by nature, inattentive, distracted, erratic, unable to concentrate for long on a single subject." It was necessary to act on this "weak and feeble will," so that, with time, he could become "stronger." A methodology would therefore have to be adopted that used "the most effective means" that led the child "to *naturally* and *spontaneously* pay full attention, creating in her/him the habit of applying himself to an activity continuously and persistently." Lima asked, "How can I grab the attention of the child?" and came up with the answer: "By observing his nature and respecting it." It was evident that the regulator of education resided in "the strength, the attention span and the resistance to tiredness of the child." The educator would be no more than an attentive student of the "vital evolution of the child."[18]

In these terms any intervention in the educational territory claiming to be an integrated solution supposed itself to be functional and technological because it was constantly structured on the *interest* of the child. It was indeed the "manifestations of the states of appearance of energies and the symptoms of the psychic states" that indicated "the only way forward in the education and teaching of a child."[19]

It is here that Lima transforms himself into an indigenous foreigner and his discourse leans on an axis constantly attributed to John Dewey: He puts forth that only through the "centre of interest of each child" can one achieve, revolve, and let loose his will. Dewey clearly became the metonymy of the discourse about the possibility of a modern kind of pedagogy. The Portuguese educator elaborated long theoretical reflections, always aiming at validating the principle that interest supposed a "reciprocal relation of convenience between the subject and the object," or, put more clearly, that interest was the "psychological form of value." The concept of interest was applied "both to the object of interest and the psychic state aroused in the subject by the presence or idea of an object." From there, he proposed a complex taxonomy of interests, based on splits such as surface–depth, form–substance, or outside–inside, whose dynamics had to be ascertained in order to overcome these opposites, which always involved acting on the second element of the pair. Lima believed that this was where, in the first instance, a battle of the saving of the self was waged, where affirmation of the conditions of possibility to master oneself was sought.[20]

It would be licit, based on these considerations, to present a law of the evolution and the succession of interests entirely based on "the Spencerian idea of progress and perfection, moving normally and naturally from the primitive homogenous to the perfectible heterogeneous (. . .). For Dewey's school and in line with Spencer's thinking," Lima explained, one had to recognize that the child's mentality undertook a series of similar states of development "to the epochs of civilisation." The individual developed fully by going through certain necessary steps that were interdependent, gradually increasing his level of interest. And if one could make a parallel between the evolution of the species and the individual, then it was necessary to discuss, as did "Dewey's school," "adaptation," and "conformity." Each of the child's states, in its evolutionary line, would correspond to the development of a certain function or aptitude whose activity causes "pleasure" because it translated "a need for adaptation" and at the same time served "as a natural step to other superior activities." In a word, the previous interests were "indispensable steps for the existence and appearance of later interests." Things happened as if all this effort of discovery of the mechanisms of subjectivation authorized—and to a certain extent was even motivated by this strategic goal—the discovery of a principle of general homogenization. The observation of the succession of interests suggested the formation of a structure of development of the child's thinking, which could be observed in at least six aspects, always aiming toward perfection: "(a) from the simple to the complex; (b) from the concrete to the abstract; (c) from passive receptiveness to spontaneity; (d) from generality to specialization; (e) from subjectivity to objectivity; (f) from immediateness to mediation in space and in time."[21] Lima argued that the school had to respect the succession of these states and adapt its entire organization to them.

The question of knowledge was presented and developed by Lima, based on the same topic of the child's interests. He argued that "the order of the subjects and knowledge corresponds to the order that individual interests appear." He championed the axioms of modern didactics—interconnection of the various contents of each branch of knowledge and their gradually increasing levels of complexity—regarding them as something that should be in perfect harmony with the "chronological and pedagogical age, the social and geographical environment of each child." The expression "active method" began to become more clear, arguing for an education that was "synthetic, united, assimilated, based not on the study of the isolated phenomena but rather on the cosmic phenomena of life"; an education now not made up of "isolated branches, but methodical units," of complexes that represented "a harmonious and organic system," of actions and of knowledge with respect to the different areas of knowledge. The psychological basis of the "active" educational processes led Lima also to speak of the "association of the states of conscience." He believed that all children were "small machines of association" and that educating them would require nothing more than organizing in them given "tendencies, associated with one another."[22]

Again, aligning himself directly with Dewey's thinking—who argued for the need for " 'times' or 'moments' of the associative process of acquisition of knowledge"—Lima explained that it was not sufficient to "construct a knowledge or a usable knowledge" and then "record it in the memory or acquire it through repetition or habit of past isolated experiences." This knowledge would be of no practical use to the learner. For knowledge to become useful, serving to be implemented and applied, it had to be associated and merged with other ideas, "constituting a constellation, a system of intellectual forces." Knowledge must be able to form "an associated group, a conscious system" that established "the unit of individual conscience in its diversity and in its infinite and kaleidoscopic variety." The ambition of the educator here was to, through the association of ideas, feelings, and sensations, format the human brain, making it more "potent and progressive" in its function of governing and relaunching individual existence.[23]

As can be seen, Lima took from Dewey an essentially "dynamic" conception of the *psy* territory. The author of *The Child and the Curriculum* appeared on the contemporary pedagogical scene as someone who had "most" furthered the "need to make education genetic." This involved no longer educating the child from outside inward, but rather from inside outward. The child should "train himself and not be trained, should instruct himself and not be crammed with knowledge built by others, should develop himself." Likewise, the icon Dewey proposed an essentially "functional" educational model based only on the child's interests. These were always viewed by him as "symptoms and signs of the deep needs," of virtualities that tend to transform themselves into "new functions" and which the educator should know how to second, placing the child in the most favorable material circumstances for their emergence. Under this functional aspect he proposed a pedagogy that aimed to develop the mental processes in their "vital biological significance, useful for present and future action." As if the psychic activities emerged as "tools" destined to promote external action and to transform individual conduct, "functional pedagogy does not forget that the child is a living being, in which one should stimulate the need for action through spontaneous means that life itself creates."[24]

The Modern School: A Microcosm of all the Methods of Social Life

According to Lima, this derived a third characteristic, which arose "as the fundamental thesis of J. Dewey: pedagogy is social." Indeed, school should accept itself as the first institution that, maintaining and respecting the natural and idiosyncratic conditions of the human being, trains the child in all the "practices of social life." School had to know how to create this environment. Lima again appropriated expressions directly from Dewey, which were given the status of redemption and salvation ideas. The school of the future would transform into a "miniature community," into a "society in which the pupils would collaborate, help one another, carry out works with a social purpose, as 'methods of life,' as 'centres of scientific curiosity, of active investigation' as starting points for the study of all other subjects. Only as such, he concluded, could the most important component of the child's soul be aroused—"the conscience of the value of work, of its social and human significance!' "

One can see how Lima identified the social as the last level in the pedagogical thinking of John Dewey, aimed at legitimizing an alternative school model. The director of the pedagogical journal *Educação Social: Revista de Pedagogia e Sociologia* could now talk like the militant who states that he has the key that enables the transformation of the most traditional structures of existence. As such, he too was ready to accept the modern principle that psychological vision should be associated to the government of men and be intermingled with it, as if "the sources of private satisfaction and human solidarity could be one and the same."[25]

Lima began to advocate a model of organization whereby the child has to be at the center of the school system and that the central goal of school is to prepare the child for active life. "The only process of preparing one for life is to live life!" In these terms a school institution of the new type had to see itself as socially structuring. The model school would therefore promote a kind of existence that led toward "upward progress," that was in constant evolution, "never being the same thing." Lima often spoke about the "Social School"—and also about "Art School," "Science School," "Workshop School"—to sum up the spirit of what his model of the reformed public school would be: an institution characterized by the plasticity and by the permanent absorption of the evolution of art, science, and work. He simultaneously idealized it as an organ and a laboratory of social life, which should include in its curriculum "all the social activities borne out of the needs felt by individuals." Each lesson should acquire the atmosphere of a studio, a laboratory, or a workshop. All the didactic materials would have to be of a technological nature and completely adaptable to everyday life. The Social School should also be based on "a free, collective and conscious pact" among all its members. "Outside coercion, violence or force from elders" was to be replaced by "persuasion, conviction, the grand principle of human solidarity, by a collective ideal to be implanted." Its pedagogical orientation, concluded Lima, would have to be of an "essential and exclusively sociological nature."[26]

He immediately felt the need to return to Dewey, to describe him this time as the most highly reputed pedagogue on the international scene, as a messenger and a prophet of the *New Discovery* of the Social School. While the philosophy and the general orientation of the Social School did not belong to him—Lima attributed the origin of the idea to the French anarchist Paul Robin (1837–1912) who put the ideal into practice in an orphanage he ran in Cempius, in France, between 1879 and 1894—there was no doubt that he embraced it completely, in the mid-1930s, in the figure of the author of *Democracy and Education*. Dewey would acquire the guise of the messianic hero:

> The pedagogical solution of the educational problem idealised and undertaken by Paul Robin, in Europe, can be seen in America personalised in John Dewey. It is he, living in a less traditionalist country, who now manages to propagate the Truth and impose it, not only in his country but also to disseminate it as the New Truth, in the name of the New World, in the Old World which, full of traditions, did not want to accept or understand this same Truth presented, then by Robin! Nobody is perfect on earth. A truth never dies. If in a given place or moment it is forgotten or does not have instigators and co-operators, in another place or moment it appears strong and imposing. This Truth has to appear, because it is necessary to correspond to an epoch, to the psycho-collective aspirations of a generation—it has to emerge, here or there, simultaneously or successively . . ., although this person is no more than the simple agent of the collective will! Therefore we can say that the Social School is based principally on the social pedagogy of John Dewey.[27]

Following this line of thought, Lima's pedagogical discourse began to converge to the theme of work. He championed work as able to produce the desired paradigmatic revolution and promote a complete education of the pupil in a different way. "For the school to be modern, whereby life is lived, functional, active, social, it has to be *of* work *through* work." In other words: Education would be carried out through work drawn up by the educator and entirely in a working environment. Indeed, only in a context of constant activity can knowledge be acquired and assimilated. The director of the Lisbon Teacher Training School never abandoned this one certainty.

Until manual educational works do not become the "basis of the entire school life of the pupil," are not effectively "the process of teaching all the subjects," do not transform into the "exclusive process of primary, physiological, hygienic and aesthetic, mental and social instruction," it is certain that school will continue to be what it has always been: "a classical artificial institution, passive, sinister and gruesomely dead!"[28]

It would be necessary to "repeat, time and again," that the activities undertaken manually constituted: (i) "a process of acquiring knowledge through the motor activity," seen as the "most in line with the psychic nature of the early years of the human being"; (ii) "a manner of refining feelings, elevating and fine-tuning the first warnings of human aesthetics"; (iii) a practical form of "transplanting social life to school, conferring it an atmosphere in which the gregarious animal is transformed into a socially conscious and sociable human being." In manual work, the effort of the child is no longer isolated and becomes "synergic. It is carried out in association, it is collective, cooperative, and common to all the pupils. It is, in a word, social, creating solidarity, the foundations of the moral and of justice." Thus with manual work, a single school practice enables the accumulation of benefits, both in the forming of personal identity and with regard to life in society. But, it is clear that the motor of disciplinary integration is always maintained in the interior of the subject. No other activity is more in accordance with the nature of the child and his "centres of interest" than manual work. This is why Lima viewed it as the "best kind of conditioning of the child's physiological, aesthetic, mental and social development." In the same vein, the major duty of the reformed school would be to furnish each subject with the conditions to experiment with the respective "manual aptitudes and to have, on the threshold of manual life, an idea of the conditions to exercise the most well-known trades."[29]

This positioning served a clear purpose of regulation of economic life: The aim of maximum profitability would be achieved when one could place, one by one, the right man in the right place. One therefore had to produce the *homos faber*, "that industrious human being, transformer and adaptor of matter for his profit," which also characterized "human nature." This is the utopian vision that directly links to a unitary image of human existence as a changing state. Past and future are interconnected by the same genealogy of perfection and progress. Lima argued for "the industrialisation of education, or rather, an education based on technology," which would be no more than a logical consequence of the application of the "biogenetic law," of the "abbreviated recapitulation," or of the "positivist theory of the epochs of civilisation." The operability of a mythical belief was accepted: The socialization of the individual, both spontaneous and systematic, "necessarily reproduces, in its large successive phases,

education of the species, both with regard to feelings and to ideas."
Therefore, the occupations to supply to the child—"a being which
lives in the state corresponding to primitive civilisation, to the primitive, manual industries, more extensive than intensive"—should not
go much further than the life of the small cottage industry or the small
manufacturing workshop. He is proud to have started in Portugal, in
his Workshop School no. 1, from 1907 onward, "a systematisation, of
integral tendency, of educational manual works" precisely in line with
this conception of civilization change.[30]

And it was precisely concerning the problem of the promotion of
the Social School that he referred to the pedagogical ideas of John
Dewey in more detail. Lima used them to argue points that he considered his own. And this meant that the American entirely fulfilled his
function of conceptual personage, heteronym, intercessor, crystal or
fold. It seemed he wanted to say: *It is not my everything when I state
it belongs to me; I am no longer I.* This is what we find in these
passages taken from the second volume of his *Metodologia*:

> J. Dewey wants the manual works to be, not mere school exercises, but
> "methods of life." They should be seen in the light of their true social
> significance. "As processes that help society to continue its forward
> march; as a means to give the child some essential notions of community
> life; as new solutions given to satisfy the needs for growing intuition and
> talent in the human being; in a word, as factors that allow the school
> itself to take on an authentic form of active social life." Manual school
> occupations are not limited, as many claim, to "knowledge of a certain
> number of practical recipes." They are aimed at the creation of active
> centres through the development of scientific investigation, on the
> materials and processes of manufacturing, and also establish starting
> points from where children may undertake the historical development
> of the human being. It is not surprising therefore, that Dewey intro-
> duces into the schools the "historical trades" and gives a leading role to
> "rediscovery" and "reinvention." Indeed, Dewey argues the following
> doctrine: "the evolution of the child under the influence of education
> goes through the phases of formation that characterise humanity . . .
> The instincts of primitive men are found in the child, whose interest is
> vividly excited by viewing the facts and manipulating things in the
> environment in which our primitive ancestors lived. Occupations sensi-
> bly selected arouse children's instincts and channel them towards
> modern aspects, making them go through phases of conscience that a
> superior race passed through to arrive at its civilised state. This is the
> purpose of the educational method through school, as states professor
> Dewey. It is the biogenetic law applied to education.[31]

According to Lima only these educational practices could give body to the pedagogical ideal of modernity, that "all education should be genetic, functional, social and sociological." With its full institutionalization, the principle of personal fulfillment would become universal. On the day that the school took the states of individual conscience as the starting point, accepting that the only valid knowledge was that extracted from things themselves, from the fingers and hands of children, the need for each person to perform a function in society that is most in line with his vocation would be realized. Only this kind of citizen, professionally fulfilled, can benefit without constraints "from his rights and practise all the duties that are characteristic of the human species." Given that it overcame the classic distinction of the productive system between thinking and doing, only the Social School was able to "reveal the skills," indicate the profession, and turn every subject into an authentic "social conscious value, eminently social and just." It would lead to the eradication, in one clean sweep, of "slavery" and "parasitism" in labor relations. The manual worker would have a "clarified, guiding and emancipating" brain; in turn, the intellectual would have the "manual skill, the power to execute what he thinks, to transform the idea into action." "Liberty" and "social peace" can only be obtained the day that education merges the manual with the intellectual inside every individual.[32]

Self-Government and Its Institutionalization in Portugal

The principle of self-government constitutes the natural conclusion of the pedagogical discourse of Adolfo Lima. It also provided the organizational conclusion, as these ideas connected concrete practical experiences of educational practices. In several texts he reiterated the principle that only the regime of the autonomy of learners could fully release the pupil from the personal tutelage of the adult because it placed him under the supervision of his own moral conscience. Individual self-governance and independence from the social group would have exactly the same existential context. Lima argued that it was necessary that the school transform "into the society of children." The school should be organized and function in order to provide the living exercise of "emancipation and freedom, of conscious and convergent action," able to make pupils, "*through practice*, future participants and collaborators of a superior, more refined social life."

Of all the forms, the organization of autonomy was not reliant so much on the domain of the psychologist, as argued in the epoch by pedagogues such as Adolphe Ferrière, but above all, on the reproduction of political institutions.[33]

The government of the self would only be clarified in a background of rules of government applied to the whole group. Personal virtue would not survive without the existence of an institutional background of social self-regulation. Indeed, Lima believed it was a question, with the formalization of self-government, of constructing an equal democratic regime within the walls of the school, so that autonomy was identified with a kind of organization in which free citizens obeyed only the laws to which they would directly or indirectly contribute. The Social School would have to be organized and begin functioning in order that the pupils exercised "an ideal of emancipation and liberty." In a "liberating education," he clarified, "there are no powers and each child has a social function to perform to contribute to the overall wellbeing."[34]

Lima was directly associated to the first school experiences of self-government in Portugal. An analysis of his biography shows that he strove to create a regime of autonomy in all the schools he worked in from 1910 onward. In February that year an Association of Pupils was created in his School-Workshop no. 1, which organized school and post-school instruction for pupils in the areas of science, art, and sport. It was, at the time, the "only organisation of its kind" in Portugal, in that it was "eminently humanitarian" and also "perfectly autonomous," because it was run by "democratically elected" pupils and teachers. From 1911 onward this model of the School-Workshop no. 1 was adopted. As a teacher of the Pedro Nunes Secondary School, Lima was one of its driving forces. This School Association had the motto, "WE EDUCATE OURSELVES," which involved nothing other than "an entire programme of social autonomy." It was organized through a federal structure, and thus there was active participation of the partners in all the decisions and structures of the organization.[35] Its various sections, not restricted to culture and sport, but also including assemblies and social cooperatives, transformed the Pedro Nunes Secondary School into a model that, from 1931 onward, spread to the state secondary schools in Portugal. Indeed, all Portuguese secondary schools were now compelled by law to create a School Association that would organize the set of activities that provided the rounded instruction of the pupil outside the classroom.

The self-government regime carried with it the most effective technology of power in the socialization of children and the young. This made it extremely welcome by the Portuguese educational authorities, regardless of their politics. Without purposely setting out to do so, Lima had imagined and tested in practice, a form of socialization that would reveal itself to be of the utmost utility for educators with completely opposite views of the world. It is very important to note that these experiences of autonomy began to be presented in the background of the monarchic regime, which was in force in Portugal until October 5, and continued in the succeeding political solution, a democratic Republic, that governed the country until 1926. But it was in the long period from that year until 1974, when the Portuguese state was defined as conservative and authoritarian, that the experiment initiated by Lima was no longer an isolated experiment. One could therefore conclude that the principle of self-government in the school, although linked to the democratic field, was indifferent to the political order. It is indeed true that its institutionalization coincided with the phase of the greatest radicalization of Portuguese authoritarianism, at the start of the 1930s, at a time when the Lisbon government aligned itself politically with Berlin and Rome, and did not condemn fascism. It was the same political regime that had arrested Lima in 1927 because of his sympathies with anarchism and with the working-class movement, which saw the utility of a moral in action that Lima began to apply in the School-Workshop no. 1. This school had been in turn the pride of the Portuguese republican and democratic regime.

It must be pointed out one final time that it was in Dewey that Adolfo Lima found this promise of a solution to the social problems using the regime of self-government:

"When School makes each child the member of a small community," states J. Dewey, "saturating him with the spirit of sacrifice, and providing him with the tools needed for effective self-direction, we will have the most profound and best guarantee of one day obtaining a dignified, loving and harmonious society."[36]

Final Considerations

This chapter aimed to show how the pedagogical thinking of Adolfo Lima incorporated the essential theses of the New Education movement, and attempted to make them compatible with the dynamics of the government of men. Always based on the heritage of Dewey, his writing began to suggest the possibility of a political subjectivation that associated the axiom of childhood interest with the institutional practices of framing of the school population.

But this association between the psycho-pedagogy and the exercising of political power in the discourse of Lima took the form of a pros- thesis in the modern educational discourse and led him to argue for the need for a new expert in the educational field, the sociologist. He spoke always as someone who took the modern *psy* vision on board but who wanted to radicalize and re-dimension it. The Portuguese educator certainly (i) shared the idea that the regimes of mental activ- ity, described by differential psychology, overcame the rationalism of the theory of the faculties of the soul; (ii) accepted that the principle of the psychic economy would have to lead to forms of education adapted to the characteristics and interests of each pupil; (iii) accepted the thesis that the whole teaching–learning process would have to be active. But to fully overhaul the traditional model the school would have to take on an entirely different nature than what was usual. It would have to be an institution of social integration and as such no longer a differentiated institution. The child would live in a miniature society and the two major guidelines of the school of the future would be work and civic participation. It was in the name of this vision, which enormously widened the universe of learning, that Lima felt it necessary to show how the research and the important solution pro- posed by New Education were, despite everything, insufficient for the effective undertaking of the Copernican revolution. Effectively, it seemed to him that the pedagogical renovation was still too caught up in the individual-genetic psychology vocabularies, in the idea that change to the educational paradigm would only be achieved through knowledge of the psychic processes. These regimes of mental activity, while allowing for the meticulous description of the child's subjectiv- ity and claiming statistical knowledge of the psychological character- istics of the whole school population, were not sufficient to operate the intended structural transformation. The productivity of the col- lective bonds and pacts would have to be added to the knowledge of the self. Personal conscience would be of no use without a social conscience.

The problem was to be solved, above all, from an organizational aspect: it was not sufficient to propose tailor-made education, adapt- ing to the idiosyncratic characteristics of each child. He would only occupy the center of the whole of school activity when he was effectively placed in a "free social regime." If the alternative model had not yet emerged in a clear form, this was due to "the need for a social ideal and the predominance of educators who were more psychologists than sociologists." Lima accepted that it was important to follow the advice of the "psycho-pedagogical differential," but for the educational regeneration to be complete "a pedagogy exclusively based on psychology should be followed by pedagogy of sociological preparation and ideals."[37]

Throughout the discursive formation of Adolfo Lima, John Dewey completely fulfilled his essential mission as the indigenous foreigner, given that, based on him, the Portuguese gradually created an invisible fabric for an educational discourse that always claimed its systemic organicism. As Popkewitz points out, "foreign names or concepts no longer exist as outsiders but with an indigenous quality that erases any alien qualities."[38] And it was as if the place and the global were entwining and interconnecting in the development of a production that, in the end, we learn should be a production of power. Indeed, Lima was able to aggregate the theses of New Education to the prosthesis, specific to the field of the political battle, according to which freedom was, before all else, a question of social conscience. All this rationalization that overcame antonyms and united opposites was present in Dewey. Whenever he tackled the topic of the moral in the educational territory, he endeavored to destroy the opposites between "interior and exterior," "spiritual and material," "body and soul," "ends and means," "duty and interest," "intelligence and character," or "social and moral." If the fundamental relation was always between the action and its consequences, then Dewey never failed to show that "moral is all education that develops the capacity to take part effectively in social life."[39] He therefore opened a new path to another design of the spiritual "nature" of the child and to the universalization of the forms of social exercising of political power completely based on the principle of self-regulation.

The pedagogical thinking of Adolfo Lima conceived, legitimized, and demonstrated the efficacy of a logic of government, and inverted the approach of the individual moral and civism in the school. These, in the institutional context of self-government, were no longer presented as a mere declaration of intent, as an ideology, to be conceived of only as practices marked by a collective pact. The expression Social Education described the constant need of the modern school to develop as an organism, with various apparatus and organs, whose fabric would be made up entirely of children. It was exactly this association from the *bodies* of the state to the subject that allowed Lima to build the scenario that would operate that utopia of the modern school. In it, the psychological domain associated itself to the social and the government of the self with the government of all. In fact, "The modern school child is the person who learns to be a 'citizen,' who has abstract responsibilities related to the governing of the state, who has 'potential' as a worker, who learns cultural skills and sensitivities for future 'use', and who is 'self' monitoring in affective and cognitive development."[40] No political power, from the radical left to the most conservative right, dispensed in the twentieth century with the temptation to reproduce the institutions that each pupil would have to conquer, on his own, when he entered in the society of adults.

In these terms the most correct association to be established is not so much between the pedagogies of autonomy and the different forms of political power, but, essentially, between the former and the same project of modernity, translated into the grand metaphor of the traveling libraries of New Education: the never-to-be abandoned hypothesis of a reversible dynamic of power-knowledge-want in the socialization of children and adolescents. It was the possibility of this moral in action that marked the appearance in Portugal of John Dewey as the conceptual personage of Adolfo Lima, who worked tirelessly to show that each singularity should be seen as a point of passage laden with principles and forces of power.

Notes

1. A. Candeias, *Educar de outra forma: A escola oficina n°1 de Lisboa 1905–1930* (Lisboa, Instituto de Inovação Educacional, 1994); A. Candeias, "Adolfo Lima," in *Educadores Portugueses*, ed. António Nóvoa (Porto: Edições Asa, 2003).
2. António Nóvoa, "Uma educação que se diz nova," in *Sobre a Educação Nova: Cartas de Adolfo Lima a Álvaro Viana de Lemos (1923–1941)*, ed. António Nóvoa and A. Candeias (Lisboa: Educa, 1995), 37.
3. Indeed, Dewey's presence in Portugal was extremely shallow and sporadic until the end of the 1960s. This is exemplified by the following two examples. One of the most important pedagogical innovation journals of the time in Portugal, the *Revista Escolar*, published regularly between 1921 and 1935, made only eighteen references to the name of Dewey, and focused solely on him in only one article; the first Portuguese translation of a text from the American philosopher is dated 1971. Up until then only Brazilian translations had circulated. L. M. Carvalho and J. Cordeiro, *Brasil-Portugal nos circuitos do discurso pedagógico especializado (1920–1935)* (Lisboa: Educa, 2002).

4. Gilles Deleuze and Félix Guattari, *O que é filosofia?* trans. Maragarida Barahona and António Guerreiro (Lisboa: Editorial Presença, 1992/1991), 59.

5. Thomas S. Popkewitz, "Globalization/regionalization, knowledge, and the educational practices: Some notes on comparative strategies for educational research," in *Educational Knowledge: Changing Relationships between the State, Civil Society and the Educational Community*, ed. Thomas S. Popkewitz (New York: State University of New York Press, 2000), 10.

6. Adolfo Lima, *Metodologia: Lições de metodologia professadas na Escola Normal Primária de Lisboa* (Lisboa: Livraria Férin, 1927), 292–293.

7. Adolfo Lima, "A autonomia dos educandos e as associações escolares: As solidárias," *Educação Social* 31–32, no. 2 (1925): 106–107.

8. Lima, "A autonomia dos educandos," 108.

9. Lima, "A autonomia dos educandos," 108.

10. Adolfo Lima, "A escola única," *Educação Social* 25–26, no. 2 (1925): 47–48.

11. Adolfo Lima, *Pedagogia sociológica* (Lisboa: Couto Martins, 1936), 299.

12. Lima, *Pedagogia sociológica*, 298.

13. Jorge Ramos do Ó, "The disciplinary terrains of soul and self-government in the first map of the educational sciences," in *Beyond Empiricism: On Criteria for Educational Research*, ed. P. Smeyers and M. Depaepe (Leuven: Leuven University Press, 2003), 105–116.

14. Henri Marion, "Psychologie," in *Dictionnaire de pédagogie et d'instruction primaire*, ed. F. Buisson (Paris: Librairie Hachette, 1882), 1761.

15. G. Compayré, "Facultés de l'âme," in *Dictionnaire de pédagogie et d'instruction primaire*, 986.

16. Lima, *Metodologia*, 292.

17. Lima, *Metodologia*, 42–47.

18. Lima, *Metodologia*, 290–291.

19. Lima, *Metodologia*, 211.

20. Lima, *Metodologia*, 296–299.

21. Lima, *Metodologia*, 308–309.

22. Adolfo Lima, *Metodologia* (Lisboa: Livraria Férin, 1932), 25–49.

23. Lima, *Metodologia*, 1932, 49–57.

24. Adolfo Lima, "John Dewey," *Educação Social* 24, no. 1 (1924): 426–427.

25. Richard Rorty, *Contingência, ironia e solidariedade*, trans. Nuno Fonseca (Lisboa: Editorial Presença, 1994/1989), 15.

26. Lima, *Pedagogia sociológica*, 281–301.

27. Lima, *Pedagogia sociológica*, 281–282.

28. Lima, *Metodologia*, 1932: 336.

29. Lima, *Metodologia*, 1932: 336–341.

30. Adolfo Lima, "A escola única," 50–53.

31. Lima, *Metodologia*, 330–331.

32. Lima, "A escola única," 58.

33. A. Ferrière, *L'autonomie des écoliers: L'art de former des citoyens pour la nation et pour l'humanité* (Neuchatel: Delachaux & Niestlé, 1921).

34. Lima, *Pedagogia sociológica*, 307.

35. When he became headmaster in 1918 of the most important teacher training school for primary schooling in Portugal, Adolfo Lima immediately created a School Association based on the same principles.

36. Lima, *Pedagogia sociológica*, 304.

37. Lima, *Pedagogia sociológica*, 303.

38. Popkewitz, "Globalization/regionalization," 10.

39. John Dewey, *Democracia e educação*, trans. Godofredo Rangel and Anísio Teixeira (São Paulo: Companhia Editora Nacional, 1936/1916), 442–439.

40. Thomas S. Popkewitz, "The production of reason and power: Curriculum history and intellectual traditions," *Journal of Curriculum Studies* 29, no. 2 (1997): 134–135.

III

European Spaces: The Southern Eastern Tiers

6

Balkanizing John Dewey

Noah W. Sobe

A Yugoslav pedagogue reporting on the advanced state of schooling in Czechoslovakia in 1934 had this to say about Czechoslovak educational literature:

> In their reviews, books and lectures many foreign pedagogues are mentioned. One thing is characteristic, however: German pedagogues are mentioned by far the least, regardless of whether they are Austrian or "Reichsdeutsche." In place of this, they emphasize Tolstoy, J. Dewey, Spencer, M. Montessori and others.[1]

Given the concerns of this book, it will come as no surprise that the mention of Dewey here is my foresmost interest. Salih Ljubunčić, the Zagreb professor who included Dewey in this list of educational thinkers, considered it extremely praiseworthy that Czechoslovaks were turning to these thinkers and not to Germans. That they were turning to foreign pedagogues was itself also something to be admired about Czechoslovak educators in Ljubunčić's view. John Dewey was one of the foreigners who could be mentioned in this eighty-page Yugoslav book on Czechoslovak schooling as harbingers of modernity. In this brief statement that puts "J. Dewey" alongside a set of other noted figures, one can almost literally envision a bookshelf in the traveling library that circulated and reassembled Dewey in Yugoslavia.

The "Balkanizing John Dewey" referred to in the title of this chapter uses the term "balkanizing" in a deliberately ironic way. In English, the word "balkanize" frequently refers to the unraveling of some entity into various subparts, akin to the twentieth-century historical pattern of countries on the Balkan peninsula of Southeastern Europe disintegrating into many smaller ethnic and religious units. Contrary to this, however, my use of the term "balkanizing" refers to one specific localization of Dewey's works. It is not meant to suggest a dilution or splintering of Dewey's ideas but rather the particular hybridized assemblage of discursive practices through which Dewey traveled in part of the Balkans, namely Yugoslavia, in the 1920s and 1930s. Instead of being a divider, the balkanization of Dewey was actually more of an assembling of multiple parts through which intelligibility was created.

That Dewey was mentioned by a Yugoslav professor of pedagogy as indicating how Czechoslovaks were wisely turning away from Germans clearly shows how the assembling of Dewey in Yugoslavia—or, as just outlined, his "balkanization"—took place in a field of multiple cultural relations. A Pan-Slavic solidarity with its Germanic other was one of the major features of the grid through which Dewey was reasoned about by Yugoslavs. The works of "J. Dewey" provide an entry point for examining the constitution of a "Slavic modernity" in Yugoslavia and the kinds of actions, knowledges, and reasoning that formed modern modes of living.

The Yugoslav balkanization of Dewey in the 1920s and 1930s engendered a multiple of modernity in which—contrary to the Weberian thesis—enchantments and disenchantments were concurrent gestures.[2] Dewey entered into the making of a Slavic modernity through ideas of action and agency that circulated with him. On the one hand, Dewey traveled in Yugoslavia as a thinker and an actor, as the exemplary embodiment of a modern, desacralized human agency. This disenchanted agency can be seen as the relocation into society of an agency once considered to reside in transcendental, divine, or natural forces.[3] This was a cultural conception of human agency that traveled in the way that Dewey was seen not just as a thinker but as a thinker who put thought into action. Yet, on the other hand, traveling in Yugoslavia with Dewey was an enchantment of action and activity, namely an enchanting of the school-based "work" activities of the child. Students' work and activities were theorized as related to a kind of "genetics," which was less a physiological/biological notion than it was the deferred location that housed the motivating, inspiring features that imbued the inclinations, dispositions, and *interests* of the child with purposive-ness. In Yugoslavia it was with the idea of the "Slavic soul" that what Weber referred to as "mysterious, incalculable forces"[4] entered into thinking about the interests and activities of the child. The soul was the elusive target of modern, twentieth-century progressive pedagogy,[5] yet the soul never achieved the kind of calculability that Weber claimed was modernity's means of mastery and was responsible for giving rise to a disenchanted world. What is worthy of note is that instead of portending a counter- or antimodernity, the enchantments that traveled with Dewey in Yugoslavia more accurately appear to have been part of the making of a modernity.

This chapter first discusses the circulation of Dewey as a conceptual persona. An investigation of the various ways Dewey was positioned allows for a discussion of how Dewey was woven into the cultural forms of modernity in Yugoslavia. The second section of the chapter examines the translations of Dewey's work into Serbo-Croatian in order to specify further the particular local assemblings of this "internationally renowned" American pragmatist and what ideas about action, knowledge, and reason circulated in the libraries through which his works traveled. These twinned lines of inquiry map the configuration of discourses and practices that made Dewey comprehensible along the particular, local enchanted and disenchanted contours taken by "modern" modes of living and the "modern self" in interwar Yugoslavia.

The Persona of Dewey in Yugoslavia

The listing of Dewey seen above put him in the midst of a crowd of thinkers of similar iconic status. In the early twentieth century and beyond, such individuals regularly circulated as conceptual persona, peopling various pedagogic literatures with international figures around whom a common grounding for modernization projects could be based. The "individuals" appearing in such lists were particular local figures not merely simulacra of an "original." References to thinkers en masse as seen in the alleged Czechoslovak emphasis on "Tolstoy, J. Dewey, Spencer, M. Montessori and others" perform a credentializing function of bringing certain global figures into particular local relations. It would be a mistake, however, to view this phenomenon merely as semiotic play involving only the manipulation of symbols empty of substance. While it is likely that similar listings can be found in numerous settings around the globe, there is a specificity to this particular listing and it proves possible to excavate historically the particular "J. Dewey" one encounters in this 1934 Yugoslav text. Tolstoy's presence as one of Dewey's shelf-mates is anything but haphazard; the Czechoslovak setting for this collection of "foreign pedagogues" is, similarly, not accidental. Interactions with Czechoslovakia were a central part of the modernization of the Yugoslav child, teacher, and school. This was a modernity that had a "Slavic" shape even as certain notions of agency and personhood were globalized through it.

Salih Ljubunčić's book on schooling and education in Czechoslovakia was based on a considerable amount of exposure to Czechoslovak education, but most immediately it grew out of a two-week study-tour of the country. In 1933 Ljubunčić led a group of thirty Yugoslav teachers on an excursion that took the group from Zagreb, through Vienna, to Bratislava and then through parts of Czechoslovakia (Žlín, Brno, Pardubica, Hradec Kralovy, and Prague) where noted educational innovations could be met with.[6] A prejudice against German educational influences pervades his text and appears in the very itinerary of the study-tour. As they were passing through Vienna, the group visited schools established for Czechoslovaks living in the city. In a report published in a Slovenian language teachers' journal one of the participants on Ljubunčić's 1933 study-tour described visiting the Comenius school in Vienna and noted, "to begin with we felt a Slavic hospitality in the middle of this foreign existence, it warmed us."[7] Not only was this school a place where the Yugoslav visitors were made to feel particularly welcome, it was a school they considered one of the most modern educational facilities in the city. The advancedness of Czechoslovak schools in Vienna, the former Habsburg imperial capital, was for Yugoslavs a testament to national perseverance.

The Viennese itinerary of this Yugoslav study-tour points to one of the conditions that made for the particularly strong Yugoslav interest in Czechoslovakia, which was that the two countries were understood as sharing a similar "historical destiny." They were newly independent "young" nations that had partly emerged out of the dismantling of the Austro-Hungarian empire at the end of World War I. This moment was frequently referred to in Yugoslav literature as a restoration of the independence that had been denied during the period of Habsburg—and additionally for Yugoslavs Ottoman—control. In the introduction to his Czechoslovak study Ljubunčić faulted Yugoslavs for "not looking beyond the borders of Austrian and German pedagogy." With their liberation Yugoslavs should now be "interested in Romance (French and Italian) and Anglo-Saxon (English and American) pedagogy," he wrote.[8] Czechoslovakia had successfully freed itself from these influences and was properly enjoying its national independence as was indicated by this particular bookshelf on which "J. Dewey" was found in 1934.

Alongside the shared temporality of both peoples living in a moment of national liberation or a postcolonial present, Czechoslovaks and

Yugoslavs shared a set of "Slavic" affiliations. Serbo-Croatian and Czech and Slovak are linguistically related languages, part of a Slavic language group that also includes Polish, Russian, Bulgarian, and Byelorussian. Slavic commonalities helped, for example, to make Czechoslovakia the most popular travel destination for Yugoslavs in the interwar era. Czechoslovaks were "our northern brothers" and Czechoslovakia was seen as the "most advanced Slavic country."

This cultural construction of Slavic relatedness smoothed the way for Czechoslovakia to appear to Yugoslavs as an attainable model of modernity. In his text on Czechoslovak education, Ljubunčić noted the pronounced American influence on Czechoslovak pedagogues such as Vaclav Přihoda, Stanislav Vrána, and Jan Úher. He even went as far as to suggest that some of the advances found in America could be channeled to Yugoslavia by looking toward Czechoslovakia. A review of the book in *Učitelj* (Teacher), the most prominent Yugoslav educational journal of the interwar era, noted that in his conclusion,

> once again Mr. Ljubunčić notes the "Slavic characteristics, or even more specifically, the Czechoslovak characteristics of this entire movement, which, in truth, bears the influences of American pedagogues." In addition, the writer hopes that "Slavic humanism will spiritualize American (and Czech) practicality, and that the reformed Czechoslovak school will be above all else Czechoslovak."[9]

The Slavic was considered, as one can see here, to have the potential to preserve a degree of local legitimacy and national authenticity in the face of modern, American influences in the sphere of education. Yet, there is the clear intimation in Ljubunčić's comment that the Slavic even has something to offer America itself. The enchantment that is suggested in the idea of Slavic humanism having the potential to "spiritualize" America is allusive to other contemporary discussions of the Slavic "soul" or "spirit" as a guiding, motivating feature of reality. The spiritualization of American pedagogy invoked here can be seen as a proposal for the enchanting of educational reform; it was a gesture toward ultimately grounding and enshrining the ontological and epistemic foundations of activity and thought in spiritual enchantments that the world offered.

The Yugoslav channeling of "J. Dewey" through Czechoslovakia was not an errant gesture. The bookshelf being examined here was one on which Dewey was assembled together with a commitment to Slavic ethnic distinctiveness. The practicality of action that Dewey's work portended had been reworked in accordance with "Slavic characteristics." It should not, however, be assumed that these characteristics were static cultural notions that preceded Ljubunčić's text in an a priori manner. It is quite evident in Ljubunčić's writings—for example, in the description of Slavic hospitality—that the "Slavic relatedness" that warranted the compatibility of Czechoslovakia as a model for Yugoslavs to study was also being culturally constructed through these very travels, interactions, and texts. Ljubunčić's positioning of Dewey as a meritorious influence in Czechoslovakia was connected to the imagining of the Slavic as a form of enchantment. This connection with the Slavic meant not only that Dewey was assembled in relation to local/regional traditions but that the balkanized Dewey actually had the potential to support and sustain the fabrication of traditions as a source of enchantment.

The cultural importance of the Slavic in interwar Yugoslavia helps to explain the placement of "J. Dewey" next to "Tolstoy." The Russian novelist and thinker Leo Tolstoy had run experimental schools in the 1890s, the most famous of which was Yasnaya Polyana where he developed "student-directed" programs[10] that Yugoslav writers often referred to as "free schooling." In the interwar era, Yugoslav educators claimed Tolstoy as an important Slavic forefather. Scholars such as Sergei Hessen (a Russian émigré who was based at various times at German, Czech, and Polish universities and whose comparative education works on Soviet and American schools were widely circulated in Yugoslavia in the 1920s and 1930s) made note of the considerable similarities between the two men.[11] Beyond their shared commitment to running experimental schools, the focus in Dewey's writings on the "life" of the child, and on schooling as a mode of living could be seen as bearing a close resemblance to Tolstoy's notion of "life." Putting Tolstoy on the same bookshelf as Spencer, Montessori, and Dewey brought Tolstoy into communication with, and put him in the same class as other persona who circulated internationally. The presence of a Slavic hero on the bookshelf with Dewey was yet another way that a cultural construction of the Slavic was interwoven into the assembling of the balkanized Dewey.

Though Dewey on Ljubunčić's list of Czechoslovakia's foreign pedagogues was pointedly unaccompanied by any German writers, this wasn't true in every case. In other Yugoslav literature Dewey was closely associated with Georg Kerschensteiner. Typically, however, Dewey was presented as the senior figure in the relationship. A book review announcing the 1935 Serbo-Croatian publication of two of

Dewey's essays in a book titled *School and Society*[12] noted that Dewey was,

> one of the world's most well known and most popular writers on contemporary pedagogical theory. Kerschensteiner, Dewey and A. Ferrière are three of the most noted and most popular Euro-American writers, theoreticians and propagators of the active school.[13]

In this linking of Kerschensteiner, Dewey, and Adolphe Ferrière activity and agency were the key points of intersection. All three were seen as "theoreticians and propagators," a characterization that hints at a modern conceptualization of agency in which thought tied to action located in people a set of functions and responsibilities that accompanied the enhanced status that individuals were able to possess as "agents of higher principles."[14] Kerschensteiner, Dewey, and Ferrière executed these functions and responsibilities in an exemplary manner. A certain disenchantment can be seen to be traveling on this bookshelf of educational reformers who were both thinkers and actors inasmuch as an ability to act, speak, and make recommendations in the "interests" of advancing society has been lodged in these individuals as opposed to in a transcendental, divine, or natural site.

However, the "activity" of the child was not solely that of a disenchanted modern agency. This Yugoslav text represented Dewey's theorization of activity and took it to be the central strand uniting him with Kerschensteiner and Ferrière. Dewey was credited with characterizing the active child's activities as having four aspects. It was recorded that Dewey thought of the "child's activity" (1) as a means for self-expression; and (2) as a way of directing the child to the satisfaction that could come from his or her own curiosity—one can note that these first two features theorized a modern self who was constituted through its own actions and out of a purposive-ness seen, tautologically perhaps, to reside "within." The 1935 text also recorded that Dewey thought that the "child's activity" was (3) a means to keep the child constantly doing work, a desideratum because "in children, thinking and doing were still undivided"; and (4) a means of forming an artistic sense. In this theorization "activity" was cast as something of a guarantee, a reliable foundation on which the individual (as well as social institutions) could be based. "Because of this," the Yugoslav reviewer wrote in reference to Dewey's characterization, "the school must educate and direct instruction according to the principle of 'teaching through work.'"[15] Connecting "activity" and "work" through

Dewey meant that the Kerschensteiner's *arbeitsschule* [work school] ideas[16] could be cast as deeply indebted to Dewey. Through this, the child's school "work," particularly when that entailed manual activities, was imbued with the redemptive qualities of a "rooted," purposeful self-expression tied to an individual practical action that could generate social progress.

In Yugoslavia in the 1920s and 1930s, the "work school" movement was arguably one of the most institutionally successful "New Education" or "Progressive Education" reform currents. A Yugoslav association published the journal *Radna Škola* (The Work School),[17] in which—as in Kerschensteiner's writings—"work" was viewed not only as vocational handiwork but as a pedagogic conceptualization of independence and self-reliance.[18] Dewey's formative influence on this movement was noted in a 1926 article from *Radna Škola*. In a report on "The Old and New School According to Dewey" the assertion appeared that "the Pedagogic influence of Dewey is powerfully felt in England and in Germany, particularly in Kerschensteiner."[19] In Yugoslav literature, these Dewey–Kerschensteiner attributions served to undermine somewhat Kerschensteiner's originality. This occurred in the two instances just mentioned when Dewey and Kerschensteiner were placed side by side one another. The assertion of considerable indebtedness also figures in a lengthy overview of Dewey's work that appeared in 1934 in *Učitelj*. The article noted that "the great German pedagogue, and in truth the greatest European Pedagogue, *Kerschensteiner* received the inspiration for his famous *Theory of Education* basically from John Dewey" [emphasis in original].[20] In the flows and networks that wove Dewey together with Kerschensteiner in Yugoslavia, the American philosopher was consistently given the upper hand. Dewey's vision of activity thus appeared on the Yugoslav cultural map as it was drawn into connection (as a formative influence) to the pedagogical theorization of work and manual or vocational education.

When Czechoslovak pedagogues turned to America, Yugoslav writers emphasized the extent to which "America" was indigenized and reworked to accord with the "Czechoslovak" and the "Slavic." This did not occur in instances when Yugoslavs remarked on the ways that German pedagogues turned to America. As presented by Yugoslavs, the Dewey that traveled on the same bookshelves as Georg Kerschensteiner had not been Germanized. Instead, the cultural representation was of a German pedagogical movement more being beholden to America than a reasoned appropriation of America. Nonetheless, as he circulated in different ways in each of these orbits Dewey helped to theorize "work" for Yugoslav educators through his ideas about action and its place as an ordering principle in educational thought and processes. As an indigenous foreigner in Yugoslavia, Dewey appeared on the one hand as a moderating figure whose American pragmatism could authentically explicate certain German pedagogical theories. And, at the same time he was a figure whose ideas about practicality and practical action were considered extremely conducive to "indigenization," or to what one saw earlier described as the "spiritualization" that could be hoped to occur when Czechoslovaks (and by extension Yugoslavs) brought "J. Dewey" into the "Slavic" world.

Routes to Reading Dewey in Yugoslavia

In the 1920s and 1930s, the works of the French professor of education, Edouard Claparède were particularly significant in the movement of John Dewey's writings into Yugoslavia. As with the conceptual Dewey who appeared in Yugoslavia, the Dewey that appeared in textual commentaries was also a particular local assembling. Claparède was an important part of this assembling, as probably one of the most widely circulated and influential of Dewey's interlocutors in the country. Because of his stature as a renowned professor of psychology at the University of Geneva and the founder of the Rousseau Institute, Claparède was a conceptual persona of international stature in his own right. The primary concern here, however, is with Claparède's packaging of Dewey. Claparède was a conduit for bringing Dewey's writings into Yugoslavia—an "envelope" that affected how the "contents" were read. What Claparède offered was a scheme for thinking about Dewey. This was a scheme, it can be noted, that supported the "Slavic" as a meaning-giving and action-orienting object of knowledge in the formation of Yugoslav schooling. Claparède's key contribution to making the balkanized Dewey intelligible was thinking about the enchantments and disenchantments that were caught up in the making of modern selves, modern schools, and modern modes of living.

In Yugoslavia Claparède was discussed as a noted child-study advocate and a pioneer in experimental pedagogy. He saw in Dewey's pedagogy three primary elements: in Claparède's scheme, Dewey's educational ideas were first "*genetic*," which meant that the education occurred not from outside but from within the child. Second, Dewey's pedagogy could be seen as "*functional*," which meant that activities of schooling were an instrument for spiritual unfolding that took account of the present and the future. And third, it was "*social*," which meant that it prepared the individual for a productive role in the larger society. This schematization of Dewey first appeared in the introduction to a 1913 collection of Dewey's articles translated into French.[21] Claparède's essay was then published in Serbo-Croatian in a 1918 journal and again a second time in 1920 as a pamphlet[22] in a series of publications on pedagogy edited by Milan Šević, the head of the faculty of pedagogy at the University of Belgrade. This same essay of Claparède's on Dewey reappeared in 1930 when a two-page summary was published in the journal *Učitelj*.[23] Thanks to the wide circulations and recirculations of Claparède's texts, thinking about Dewey as offering a *genetic, functional, and social* pedagogy moved into Yugoslavia.

(It can be noted that Yugoslavia was not the only Central/Eastern European Slavic country to meet Dewey through Claparède. Claparède's essay was also frequently cited in Czechoslovakia. Especially in light of the perceptions seen earlier, it is quite interesting to find that in his 1930 book on American education Jan Úher turned to explaining Dewey's educational ideas to his Czechoslovak readership at least partly through the schema presented in Claparède's essay.[24] This recourse to Claparède isn't particularly surprising except that Úher's book was expressly written after and on the basis of his study-tour travels to the United States.)

In Claparède's conceptual schema, the *genetic* aspects of Dewey's educational philosophy concerned his ideas about development and how the teacher ought to respond to the desires and interests of children. What Dewey offered, according to Claparède, was an "understanding of [the child's] interests as a genetic symptom" and a way in which "we can follow the child's nature."[25] It can be noted in passing that the notion of development Dewey elaborated, for example, in *Democracy and Education*, differs somewhat from a concept of genetic unfolding that could be nurtured and monitored by tracking the expressions of the child's interest, the main point for present purposes is that the latter idea traveled in Yugoslavia as an authoritative presentation of Dewey. Claparède's reading of Dewey, in fact, put an extremely strong emphasis on the centering of education in the child, in "genetics"—an emphasis that seems to have been keenly picked up on in Yugoslavia.

In its Serbo-Croatian translation, Claparède's essay was prefaced by an introduction written by Milan Šević that indicates how both enchantments and disenchantments were present in this thinking about education and its "genetic" dimensions. Presenting the Claparède/Dewey concepts that his readers were about to be exposed to, Šević opined, "everyone has a certain capital in their aspirations and impulses that must be pursued in order to move forward." He continued, "the problem of education is this, to discover that capital."[26] With this statement Šević tied Claparède's "genetic" categorization to a theory of progress, a "moving forward." This "capital" was something as-if hidden inside the child. Child study could reveal something of this genetic capital and a child-centered pedagogy could nourish it. Yet, although "genetic capital" was something the child was considered to possess to begin with, it was not something static. Yugoslavs read through Claparède that for Dewey the psyche was not a static system but a dynamic process. These "genetics" were, accordingly, not the feature of an a priori determinism but the proper, "natural" material for educators to work on.

In pursuing and educating by means of the child's aspirations and interests, the disenchantments of a scientific mode of reasoning and acting took shape. Through the research in the field of experimental pedagogy and at institutes such as Claparède's Rousseau Institute, education scientists could carefully study child interests and activities as symptomatic forms of the child's nature. The "natural," "genetic" entity so well lodged in the child's interior was, however, the deferred object of the science and an elusive "thing" that could be attributed with an enchanted purposiveness. In this regard, there appear to be certain connections between Claparède's ideas about the "genetic" and Henri Bergson's vitalism,[27] with the inner, genetic nature of the child serving as an "élan vital." Thanks to Claparède's categorization, the modern approach to teaching and thinking about the child's interests that circulated through Yugoslavia in affiliation with Dewey was one that constructed enchantment and disenchantment in related gestures.

By analogy, the "Slavic" could similarly be understood as an inner, motivating spirit that needed to be pursued by Slavs. Finding a "Slavic soul" and properly putting it to use was a major goal of the many Pan-Slavic meetings that were held in the 1920s and 1930s. In this period Yugoslav educators, along with sociologists, geographers, mathematicians, and even beekeepers[28] attended international Pan-Slavic conferences with Poles, Czechoslovaks, Bulgarians, and representatives from other Slavic countries. Pan-Slavism could provide a key element of the "genetics" inside the child, and, relatedly, it could provide a "genetics" for thinking about rural villages. The Pan-Slavic literature frequently valorized village/rural schooling as the source of something quite like an *élan vital*. The above mentioned Milan Šević was the chief Yugoslav delegate to the 1931 Slavic Pedagogical Congress in Poland

and in one of his reports he described a field-trip to a Polish village school near Warsaw, writing that the village preserved values and personality, "which the cities lost, not only for themselves but for the entire nation."[29] For Yugoslavs, the "genetics" of the village, of the child, and of the Slav were enchanted and guiding objects. They were capital to be pursued; they were dynamic objects that allowed movement forward. Because of its emphasis on genetics, Claparède/Deweyan pedagogy allowed for its own indigenization, and acceded to its own balkanization, on a theoretical and practical level.

The *functional* aspects of Dewey's pedagogy were consistent with a functional psychology that, for Claparède, held promise for renovating associationist psychology with an understanding of consciousness as less static and more attuned to environmental adaptability in a Darwinian sense. As mentioned above, Claparède emphasized the dynamic as opposed to static features of the psyche, which was an emphasis that corresponded to viewing the human as a whole unit with functional integrity in adapting to its needs and its environment. The activities of the child were an important concern in this respect and Claparède noted that educators needed to make sure that the child's activities and work accorded with his or her "interior needs."[30] As "functional," the child's activities thus had the potential to provide a surface for pedagogic interventions.

The connections between functional psychology, Dewey, and Kerschensteiner's work school movement were noted in a specially commissioned survey of Czechoslovak education that appeared in Serbo-Croatian in 1938. In a passage that captures many of the themes previously discussed in this chapter, a Slovak school inspector by the name of Franjo Musil explained to Yugoslav readers the historical development of Czechoslovak didactics. Musil's story was noticeably similar to the tale that Salih Ljubunčić told through the "Tolstoy, J. Dewey, Spencer, M. Montessori" bookshelf that this chapter began with. According to Musil, up until 1918 German and Austrian pedagogies had been forced on a Czechoslovakia that with independence post-1918 was able choose more freely what best suited its educational needs. The free school [*slobodna škola*] movement that emphasized "active methods" and eclipsed the "passive learning" of information (as under the old regime) was seen by Musil as forming "under the influence of studies in functional psychology which freed children to proceed through a system of activities that would develop their dispositions." This led, he noted, to the organization of schooling around "centers of children's interest." Musil's bookshelf of pedagogic leaders in this area included "Montessori, Decroly, Claparède and, in particular, Dewey." In the traveling of this library to Yugoslavia, it can be noted that Dewey's theorization of activity was taken as a key concept warranting the focus on the child's work. Musil stated that this impulse had even led to the establishment of "educational handiwork" being established as a required school subject for boys in Czechoslovakia.[31] Once again Dewey traveled in Yugoslavia in the company of Kerschensteiner; here he has also been put in the company of Claparède and functional psychology and figures as the philosopher of a social redemption promised by pedagogic work centering on the child's activity.

The *social* aspects of Dewey's pedagogy were, in Claparède's view, infused throughout Dewey's ideas. He noted that Dewey responded to contemporary societies ("the new conditions of our new civilization") when children were severed from many of the natural occupations that were once adequate to "develop a social instinct." In the essay Claparède suggested a large concern with anomie by mentioning the loss of family life, and the phenomenon of parental work outside the home meaning that children were not kept to their work or tasks. Claparède concluded that it was only the school that could fix contemporary social life. And toward this end he turned to Dewey's concept of the "the school as a small community." He also once again turned to Kerschensteiner and the notion of handiwork as "the best type of work for interpersonal cooperative work" and "work in a community"[32] as educational strategies that would redeem society.

Milan Šević, in his introduction to Claparède's piece, also took up the question of what it was that would allow people to live well together. He drew out of Claparède/Dewey the collective development of individual self-responsibility and a healthy reason as the signature features for productive social living in contemporary times. Related to this, for both Šević and Claparède, one of the things that was central about Dewey was that not only did he think such things but that he carried them out—an observation already discussed earlier. Practical actions and a restructuring of human agency were offered toward the goal of repairing and moving society forward as a successful human enterprise.

Conclusion

Balkanizing John Dewey was a cultural phenomenon occurring in a number of uncoordinated, overlapping ways. Dewey was circulated and reassembled in Yugoslavia in the 1920s and 1930s as a harbinger of modernity. This modernity, however, was one that could be flexibly molded and enchanted and disenchanted to fit local conditions. The concepts of practical action that circulated with Dewey could be Slavicized—something Yugoslavs considered their "northern brothers" the Czechoslovaks to have done quite successfully, for example, "the Czechoslovak school will be above all else Czechoslovak." This adaptability wasn't only the result of contemporary thinking about appropriation, it had also to do with the skills and dispositions necessary for living in uncertain times and shifting cultural terrains. Dewey's localization and assembling in interwar Yugoslavia had centrally to do with the assembling of one of the multiples of modernity.

On the modern bookshelves and in the libraries with which Dewey traveled he was reassembled with other conceptual persona, international figures who too each had their specific balkanized versions. Among these figures, however, Dewey most consistently appeared in Yugoslav literature as a theorist of actions and activity. The theorization of "action" attributed to Dewey was tied to a modern mode of living organized according to individual independence, self-reliance, and self-government. The modern self that could be glimpsed and fabricated through the proper pedagogical use of the child's work, activities, and interests was one who was constituted through its own actions. This way of acting was both a doing and way of thinking inasmuch as action tied to interests/curiosity/dispositions generated principles of reflection and criteria for evaluating knowledge. In the assembling of Dewey both as persona and in textual commentary on his work the "activity of the child" provided a sound basis for reliably intervening on the individual and for reconstituting the social as a domain of interaction and mutuality. Yet this activity was both enchanted with a guiding purposiveness and disenchanted in itself and in the educational reformer's progress-oriented acting according to a set of assumed functions and responsibilities.

John Dewey's active and acting child appeared in Yugoslavia as the best way that true nature could be expressed and nurtured in a modernity. The "genetics" of this child were offered up to the disenchanting lenses and tactics of scientific study that would be undertaken by psychologists and through experimental pedagogy. Yet, an enchanted "core" remained—the seat of purposive-ness, intentionality, and volition of which interests and activities were, in Claparède's word, only "symptomatic." The attribution of a spiritual mystery to the soul was a central idea that traveled with Dewey in Yugoslavia. Whether this was the "Deweyan" in a Hyde Park or Morningside Heights sense is less important than the possibilities this engendered for Dewey as well as "J. Dewey" to comply with and support the importance of the "Slavic" in forming and enchanting the modern Yugoslav child and school. The balkanization of John Dewey was a feature in the creation of a "Slavic modernity" in Yugoslavia in the 1920s and 1930s. This modernity was distinct in one respect in being so commonly articulated along a network of Yugoslav–Czechoslovak interaction and not according to a rigidly preserved center–periphery dynamic of relations. In other aspects, Yugoslavia's "Slavic modernity" was distinct in bringing together the educational reformer's agency with the agentic activities of the child. Theorizing Dewey's active and acting child through the "Slavic" meant that purposive-ness of action could be desacralized, disenchanted, and located within society—and, it could at the same time be naturalized, and re-enchanted by being located within the spiritualization of a "Slavic genetics" that was individual and social.

Alongside this, a certain pragmatism was ensconced in the very processes and activities of John Dewey's balkanization. Cultural thinking in the 1920s and 1930s about the dissolution of barriers to the flows of knowledge and the putative "universality" of social problems and their solutions generated a progressive point of view that meant that Yugoslavs were drawn into the project of pragmatically creating a universe. In this world (here it has been a "Slavic world") actions generate their own knowledges; serialistic, repeated actions reinforce and create a mental world in which these actions fit. The progressive point of view in Yugoslavia in the 1920s and 1930s was a view on a world built around doing and progressing—a world of (pragmatic) involvement in that acting and advancing.

Notes

1. Salih Ljubunčić, *Školstvo i prosvjeta u Čehoslovačkoj: s osobitim obzirom na pedagošku i školsku reformu*, ed. Salih Ljubunčić, *Biblioteka "Škole Rada"* (Zagreb: Naklada A. Brusina Naslj. V. i M. Steiner, 1934), 52–53.
2. My thinking on the twinning of "disenchantment" and "enchantment" is indebted to Jane Bennett, "The enchanted world of modernity: Paracelsus, Kant and Deleuze," *Cultural Values* 1, no. 1 (1997).
3. John W. Meyer and Ronald L. Jefferson, "The 'actors' of modern society: The cultural construction of social agency," *Sociological Theory* 18, no. 1 (2000): 101.
4. Max Weber, "Science as a vocation," in *From Max Weber: Essays in Sociology*, ed. H. H. Gerth and C. W. Mills (Oxford: Oxford University Press, 1981), 139.

5. See the discussion in Thomas S. Popkewitz, *Struggling for the Soul: The Politics of Schooling and the Construction of the Teacher* (New York: Teachers College Press, 1998).

6. Salih Ljubunčić, "Naučno putovanje naših učitelja u Čehoslovačku," *Napretka i Savremena Škole*, no. 5–10 (1933).

7. Andrej Debenak, "Vtisi iz učiteljske studijske ekskurzije po Čehoslovaški," *Učiteljski tovaris* 62, no. 10–11 (1933).

8. Ljubunčić, *Školstvo i prosvjeta u Čehoslovackoj*, 5.

9. "Salih Ljubunčić: Školstvo i prosvjeta u Čehoslovačkoj [book review]," *Učitelj* 44, no. 3 (1934): 235.

10. Bob Blaisdell, *Tolstoy as Teacher: Leo Tolstoy's Writings on Education* (New York: Teachers and Writers Collaborative, 2000).

11. Tadeusz Nowacki, "Wstęp," in *Sergiusz Hessen: Filozofia — Kultura — Wychowanie*, ed. Maria Hessenowa (Wrocław: Polska Adademia Nauk, 1973).

12. The book included two of Dewey's essays, "The School and Social Progress" and "The Child and the Curriculum."

13. "Đon Duji: Škola i društvo [book review]," *Učitelj* 50, no. 4 (1935).

14. Meyer and Jefferson, "The "actors" of modern society," 105.

15. "Đon Duji: Škola i društvo [book review]."

16. See Hermann Röhrs, "Georg Kerschensteiner," *Prospects (Paris)* 23, no. 3/4 (1993).

17. *Radna Škola* was published between 1924 and 1929. In its final years the journal was affiliated with *The New Era*, the flagship publication of the New Educational Fellowship. Unlike *The New Era*'s other international associates in the 1920s and 1930s, however, *Radna Škola* published almost no translations of *New Era* articles.

18. This position pervades most of the issues and appears in the letter from the editor (Jovan S. Jovanović), which launched the first issue. Alongside stating that one of its goals was to have "manual work" (*ručni rad*) appear as a school subject the editor noted his hope that "work" would become a "principle" from which everything in pedagogy would follow. Jovan S. Jovanović, "Našim Čitaocima," *Radna Škola* 1, no. 1 (1924).

19. Vlad. Spasić, "Stara i nova Škola po Đonu Duiju," *Radna Škola* 2, no. 5, 6, 7, 8, and 9 (1926): 127.

20. Dragoljub Branković, "Pedagogika Đona Duja," *Učitelj* 44, no. 2 (1934): 172.

21. Edouard Claparède, "Introduction," in *L'Ecole et l'enfant* ([John Dewey] Neuchatel, Paris: Delachaux & Niestlé; Librairie Fischbacher, 1913).

22. Edouard (Claparède) Klapared, *Pedagogija Đona Đuia*, trans. Milan Šević (Beograd: Izdanje Knjižarnice Rajkovića i Dukovića, 1920).

23. Z Dirić, "Pedagogika Đona Đuia—Ed. Klapared," *Učitelj* 10, no. 6 (1930).

24. Jan Úher, *Základy Americké Vychovy* (Praha: Čin, 1930), 103–106.

25. These excerpts are translated from the Serbo-Croatian. Klapared, *Pedagogija Đona Đuia*, 19.

26. Milan Šević, "Introduction," in *Edouard (Claparède) Klapared, Pedagogija Đona Đuia* (Beograd: Izdanje Knjižarnice Rajkovića i Dukovića, 1920), 4.

27. Claparède was familiar with Bergson's work Daniel Hameline, "Edouard Claparède," *Perspectices: revue trimestrielle d'éducation comparée* 23, no. 1–2 (1993): 6, Even though he disputed some of Bergson's arguments (Tröhler, chapter 3, this volume).

28. Noah W. Sobe, "Cultivating a 'Slavic modern': Yugoslav beekeeping, schooling and travel in the 1920s and 1930s," *Paedagogica Historica* 41, no. 1–2 (2005): 145–60.

29. Milan Šević, *Prvi Slovenski Pedagoški Kongres* (Beograd: Pedagogijska Knjižnica, 1932), 22.

30. Klapared, *Pedagogija Đona Đuia*, 23.

31. Franjo Musil, "Razvoj i današnje stanje Čehoslovačke didaktike," in *Pedagoška Čehslovačka, 1918–1938*, ed. M. P. Majstorović (Beograd: Izdanje Jugoslovenskog Učiteljskog Udruženja, 1938), 28.

32. Klapared, *Pedagogija Đona Đuia*, 25.

7

John Dewey's Travelings into the Project of Turkish Modernity[1]

Sabiha Bilgi and Seçkin Özsoy

Introduction

The "struggle" to catch up with, and become part of, Western civilization in Turkey has a centuries-long history. Having a Western future in Turkey, however, means taking on a national identity that also sees Turkey in terms of lack (of Western qualities) and of "being behind." The Kemalist project of the Turkish national elite was named after the military and civilian leader who sought to Europeanize the nation from the 1920s. In this project, the "West" stands as the destination point for "the nation." Turkey has been running away from a "past" since the 1920s. There is no "present" between the "Western future" and the "never passing" past. In Ahıska's words, "the present is denied in its heterogeneous experimental terms and reduced to a permanent crisis that the Turkish national elite has 'struggled' to evade from the very beginning."[2]

Unceasing numbers of educational reform proposals are released in order to close the "gap" between the present/past and the future. There has never been a shortage in the supply of reforms, seeking to better the school and its pedagogical practices in/through which the "modern/Western" way of being and living will be attained. Yet ironically the present-day educational system is in such a pathetic condition that nobody, even the minister of education, dares to talk of any progress or betterment. Current reports on the condition of the educational system testify to the fact that the school produces and reinforces the inequalities and injustices of "the present." The release of these reports, on the other hand, does not lead to a critical engagement with the present. Instead, as İnsel notes, it gives a start to a "mass mourning" for the condition of the educational system.[3] The lament is for the state of the "nation" despite its centuries-long struggle to become part of modern/Western civilization. Speaking of failure, not success, these reports are the paramount technology that breeds a peculiar way that the collective imagination constructs Turks as "the Western" and "the Other" at the same time. While projecting people into this collective narrative of "We," the practices of modernity divide and exclude people along the axis of "the West" and "the Other." Rendering individuals the objects and subjects of intensive control and indoctrination, the practices of modernity fabricate an individual who has no sense of the historicity of "the present," refusing to "know the realm of forces that produce things as they are."[4]

This chapter is about the emergence of modernity in Turkey. More specifically, the chapter examines the role of John Dewey as an "indigenous foreigner" in the constitution of modernity as a "fixed" and "thing-like" future associated with "the West." It investigates the ways in which Dewey's ideas of "individual," "community," "democracy" and "education" have been assembled/reassembled and made understandable in Turkey. The chapter aims to historicize the "present" practices of schooling that are "reduced to a permanent crisis that the Turkish national elite has 'struggled' to evade from the very beginning."[5] By looking at how Dewey's ideas were put into work "from the very beginning" the chapter aims to attend to "the present" of schooling that might bear no resemblance to what Dewey conceived as a democratic and progressive school.

There are already quite a number of studies examining the role of Dewey in the project of Turkish modernity. These studies tend to reduce the analysis of Dewey to a problematic of more/less successful imitation of the original.[6] This study differs from the previous analyses in that it rejects the conceptualization of Dewey's ideas as a fixed "thing" that is applied, imitated, adapted, and/or imposed in/on/to the "context" of Turkey.

This chapter aims to "contextualize" Dewey in Turkey by focusing on his role as "indigenous foreigner" in the "contextualizing" of Turkey. In other words, first, the chapter does not treat Turkey as if it is a given national "context" to which the "alien" ideas of Dewey were borrowed from abroad and in which their application is juxtaposed to a pre-known consequence. Rather, it pays attention to the construction of Turkey as a national locality. It looks at why and how Dewey traveled in the production of Turkey as an "imagined community,"[7] representing "an imagined state of being,"[8] both individual and collective. Second, as Popkewitz argues, we assert that Dewey is not a "person" in the conventional sense.[9] Rather, he personifies the ideas of "individual," "community," "democracy," and "education" into which "diverse life worlds and conceptual horizons about being human" are "translated."[10] As noted by Chakrabarty, the process of translation modifies "existing archives of thought and practices." What makes these ideas universal is the translation process itself. The same translation process also destabilizes the universality of these ideas and produces different discursive practices.[11] Instead of going after the tendency of the existing studies to take "difference" as a "starting point" and subscribe to some universals in the analysis of the Turkish experience of modernity in education, in this chapter, we follow the call of Gupta and Ferguson, take "difference" as "an end product" and go on to explore the production of this "difference."[12]

This chapter consists of four sections. The first section delineates the notion of "multiple modernities" upon which the analysis is built. The second section presents a discussion of the early reform movements in the late Ottoman Empire era in which the nationalizing/modernizing/westernizing Kemalist movement has its roots. The third section is devoted to the analysis of what made Dewey "travel" into the public education discussions of Kemalist elite and what cultural purposes Dewey was to serve. The final section focuses on the educational initiatives in the early years of the Republic, including Turkish Hearths, the People's Houses, and finally Village Institutes. The significance of these three different reform movements is that they involve different projects of modernizing the nation through modernizing the individual within Turkey.

Multiple Modernities?

In academic circles, societies other than the Western generally are seen as coming to the question of modernity from without. These societies are considered not just as "not-modern" but as "not yet modern." "Being modern" represents a way of life, reached through rationalism in its origin, the West, and defined by a set of adjectives, such as industrial economies, scientific technologies, democratic politics, secular worldviews, and so on. "Becoming modern" is seen as a result of the impact/influence of the West on other national locales. As a unidirectional process, modernization represents a movement that stems from the effect of the West/the global and leads people with diverse ways of living to move towards a unitary modernity. The present disjunctures in the social, economic, political, and educational arenas of countries across the world, then, come to be seen as a marker of an incomplete transformation. Following this rationale, Wolf-Gazo, for example, not surprisingly ends his analysis of Dewey in Turkey with the statement that "the applied Enlightenment has not yet completed its task."[13]

This widespread account of modernity as a "coherent and integrated whole" has been under attack recently.[14] Turning its critical gaze on the universal and evolutionary theories of historical change of societies, poststructural and postcolonial scholarship rejects the treatment of any "difference" as a marker of "incompleteness" as posited within a Western historicity. As the idea of "multiple modernities" implies, recent scholarship seeks to understand diverse ways of contemporary human experience around the world within the pluralized notion of modern. The insistence on the pluralization of modernity has nothing to do with the celebration of relativism. Neither does it mean that Western discourses of modernity have no relevance anymore. The insistence on the pluralization of modernity is concerned with understanding how different systems of values about the self come into play in the construction of modernity.

Our focus on Dewey's "travels" to Turkey is our approach to considering the issue of multiple modernities in the construction of modern selves as historical practices. Our concern is with understanding the divergences of modernity in Turkey. This is accomplished by viewing the way in which the political elite of Turkey bring into their efforts to change not only the institutions of the state and society, but also the modes of living through which individuals are to tie their individual practices and thought to collective notions of the nation. The following sections focus, first, on "the modernizing offensive" that takes place in the early decades of the twentieth century in Turkey, differentiating but also placing the reform efforts within the context of its Ottoman legacies. This social and political context provides an overlapping set of assumptions and cultural norms through which Dewey travels into the project of modernity in Turkey.

A "Modernizing Offensive": Turkish Modernization Project and Its Ottoman Distinctions

Turkish modernity can be characterized by what Peter Wagner calls, a modernizing offensive "in which a small group use their power to spread modernity into society."[15] The main themes of the Turkish project of modernity were "nation" and "Western civilization." The project promised to build an independent nation taking charge of its own destiny, to lift the nation up to the level of "Western civilization," and to build "a perfect democracy." What made this project of modernity unique is that Mustafa Kemal Atatürk and his followers called Kemalists were to "work for something [the Turkish nation] which did not exist as if it existed and to make it exist."[16] Introducing new rules and regulations for the everyday practices of people, the project aimed to make them into a citizenry with obligations and rights vis-à-vis the new established nation-state. The transformative capacity of the project proved to be impressive. It produced subjects who see themselves as part of the national community, claiming rights, such as the demands of Kurdish population for cultural and linguistic recognition. The other distinctive characteristic of the Kemalist modernization project and its ironic quality was that it was never meant to be a negotiation process. The demands were suppressed in the name of safeguarding "the indivisible integrity of the state with its territory and nation."

Contrary to widespread accounts, the Kemalist project of modernity does not constitute a sudden and total break with the Ottoman past. Neither can it be seen as the importation of "alien" Western values into the society. The Kemalist project of modernity developed out of, and built upon the centuries-old preoccupation of the Ottoman state elite with the problem of saving the state, upon which their very social position, legitimacy, and livelihood was dependent. As Kazancıgil argues, Kemalism was "the last and most successful response" to this age-old problem.[17] As "an intensification, radicalization and culmination" of the late Ottoman era reform movements that developed through a long and complex interaction between the Ottoman Empire and the West,[18] the Kemalism led to the emergence of a new political and cultural structure. The new Republic was to be quite different from the Ottoman Empire; yet it was not to represent a decisive break with the past. We explore these conditions of political culture as a way to enter into a discussion of the traveling library in which Dewey was assembled in Turkey.

If we look first at Ottoman political culture, it shared the characteristics of "patrimonialism," while still keeping its own particularities.[19] The state was ruled in the name of the dynasty. Identical with the personality of the sultan, the state was absolute. The territory of the Empire comprised a variety of ethnic, tribal, religious, and linguistic communities, and the state was very careful not to intervene in the everyday life practices of these groups. To do otherwise, for example, to introduce certain regulations and laws in education or in family, would have resulted in the emergence of a space in which the formulation and implementation of policies were negotiated by these various groups. It would have led to the formation of a social existence that would have escaped from the control of the absolute state. As İnsel notes, the long-lasting and extended authority of the state over its subjects from Balkans to the Arabic peninsula depended on the existence of these diverse groups as homogenous entities having no relationship with each other.[20]

Hindering the development of networks among these different groups, the relationship of the Ottoman state and these groups was like the authoritarian relationship of father and children in a household. The state exercised its authority over the "household" through its military and civil bureaucratic officials to whom the monopoly of vital resources, legitimacy, and authority was reserved. What the category "Turk" represented was a variety of peasant tribes in Anatolia. The name "Turk," however, meant nothing to them. The Ottoman state was theocratic. Yet this does not mean that Islam had a unitary concern, and the believers in Islam were the subjects with whom the state identifies itself with and through whom it gains its justification. Rather, through the possession of the Caliphate,[21] the state was to use Islam as a means that was beyond any societal, worldly, or human in order to justify its absolute power.[22]

The nineteenth century was a turning point for Ottoman state. The state was losing its control over the provinces. With increasing nationalist movements, various groups were proclaiming their independence, and there was a population that was completely indifferent to the dissolution of the Empire. In this period, the idea of the superiority of the European societies began to gain acceptance at the state level. Acting like an interface between the Western world and the masses, the state elite was to come to an agreement that the imitation of "Western ways" was essential in order to save the state.[23]

On the other hand, the presumed superiority of the West and accordingly what constituted "modern" and "backward" in the late Ottoman Empire era was seen as institutional and administrational.[24] Initiated with *Tanzimat Fermanı* (the Reform Edict) in 1839, the reforms marked the nineteenth century with successive attempts to modify the existing administrative, educational, military, and judicial institutions to resemble Western models. These early reform movements included the establishment of secular schools in the cities in which the state elite was trained for the implementation of new laws and regulations.[25] Aiming to save the Empire by polishing its "superstructure" in accordance to "Western" ways, these reforms had a very modest penetration into the society. These reforms were targeted for the elites, but they embodied a particular cultural orientation that can be associated with modernity—the organizing and planning of the present for the future. As Mardin argues, the statesmen of the *Tanzimat* were concerned with "shaping the present for future use."[26] Mass schooling was not to be an issue in the state agenda until the new generations of state elite acquire their image of Western societies in these secular schools, leading them to think of the existing organization of social life in the Ottoman Lands as the reason of the dissolution of the Empire and therefore in the need of modification for the future. Indigenized Dewey, then, was to be a reference for a cultural regeneration project, of which the aim would be regarded as the replacement (not even transformation) of the existing community with a different one.

The Kemalist generation was educated in these secular Western schools in the city.[27] In classrooms of orderly rows of desks and chairs and segregated from Ottoman life, they studied new knowledge—geography, physics, law, administration, and so on. What they did in these schools was textbook learning. As Mardin writes, "life as described in books was more real than life itself."[28] Attributing the decline of the Empire to the "Eastern" or "Oriental" culture that prevailed in the Ottoman lands, the Kemalist generation concluded that unless the state was renovated radically in accordance with the universal principles of citizenship, nationalism, and secularism, it was doomed to death. The Kemailists played a great role in the Orientalizing of "Turks" as backward, ignorant, fanatic, and fatalistic. The Kemalists fashioned a "Western civilization" that was based on the idea that it must be implemented as quickly as possible if the state were to survive.[29]

After the War of Independence, the Republic of Turkey was proclaimed in 1923. The borders of the new homeland were drawn, and it was declared that the bounded place is the sacred fatherland in which the Turkish nation of the new modern state lived. The form of government of the new state was to be republican, and it was stated that sovereignty belongs to the nation. By attributing the predicament of the people to the traditional and moral standings coming from the Islamic-Ottoman heritage, the existing life in the Ottoman Lands was normalized. Although the past was thought to be the root cause of its predicament, the Turkish nation also needed a "Golden Age."

Constructing the pre-Islamic Central Asian civilizations as the golden past and locating its "other" in Arabic and Persian cultures, the nationalist discourse postulated that the pure and glorious Turkish people had been corrupted by the influence of these cultures. As the true and rational will of its people, the state was now to take the responsibility for cleansing elements that corrupted its pure and glorious character and put them on track their natural drive toward being a modern, developed, and democratic society. The successive radical and not-so-radical reforms, ranging from the replacement of Arabic alphabet with the Latin to the changes in the dress code were to remove the visibility of any character that resembles the Ottoman and Islamic past, which was seen as conflicting with the progress and democracy.[30] Through their expansion across Turkey, the secular schools were to keep making the modern individual of/for the new nation-state. The new Republic was to be no more the sacred representative of God on earth but the embodiment of another sacred cause, the modernizer of Turkish people.

Dewey's Travels into the Project of Modernity

Dewey's visit to Turkey was at the invitation of Ministry of Education to offer the reports of recommendations for restructuring and reorganizing the educational system, which we will discuss later. His visit coincided with the period in which Dewey began to enjoy an international reputation. But it would be an oversimplification to regard the interest of the Republican elite in Dewey simply as a reflection of the intellectual fashions of the time.

Dewey as a Modernizing Discourse

First, for the state elite, the traditional and religious moral values of people were the main cause of their predicament. The insistence on the link between science, morality, social order, and betterment in Dewey's texts were to provide the elite with an objective basis for their project of converting the masses of Anatolia to a new secular religion. The Republic was to be the new religion, from which all moral principles will be drawn. As Fay Kirby Berkes argues, the elite were:

> . . . concerned primarily with the highest problems of moral transformation. They were moralists who had noted the failure of the traditional values, religious and ethical, to exercise effective control over the Faithful under the new social, economic, and political conditions. For

them, new and secular conceptions of the Good, the True, and the Beautiful would be guides to action, the only valid creation of these would be individual Reason.[31]

Second, Dewey's thought that the school is the primary institution to project the population into a citizenry sharing a set of worldly principles and values fell on receptive ears among the Republican elite. Dewey was to be a name invoked in the Turkish modernization discourse, assigning the school to the task of "teaching" the requirements of the new regime and democracy in its sterilized environment to transform the outside.

Third, Dewey's ideas of "public interest," "common good," and "social organism" were captivating for the Turkish Republican elite who imagined a national community characterized by harmony instead of dissent and struggle. These ideas were translated into the modernization discourse as "the interest of the nation," "the common good of the nation," and "the nation as a social organism." The Turkish nation was to be a "Great Community" of the individuals who work cooperatively for the sake of the sublime principles and purposes of the nation's modernization and development. Dewey's writings were connected to the institutions of centralized national education that were to be assigned to the role of making the individual conscious of his/her capacity in regulating his/her conduct where the rights of the individual were to conform with the founding principles and aspirations of the nation-state. In other words, there was no "individual" outside and before the realm of the nation and its sublime purposes that linked the pursuit of individual self-realization, individual good and perfection with the norms and values ascribed to the common good and perfection of the nation.

Finally, with his reports on, and recommendations for, the Turkish education system alone, Dewey was to give the modernizing elite a justification for their intended reforms and to bless their mission to civilize the uncivilized masses for the future democracy, which they would later join. Dewey began his report by drawing attention to the necessity of setting down the main aim and purpose of the Turkish schools:

> Fortunately, there is no difficulty in stating the main end to be secured by the education system of Turkey. It is development of Turkey as a vital, free, independent and lay republic in full membership in the circle of civilized states. To achieve this end the schools must (1) form proper political habits and ideas; (2) foster the various forms of political habits and ability, and (3) develop the traits and dispositions of character,

intellectual and moral which fit men and women of self-government, economic self-support and industrial progress; namely initiative and inventiveness, independence of judgment, ability to think scientifically and to cooperate for common purposes socially. To realize these ends, the mass of citizens must be educated for intellectual participation in the political, economic and cultural growth of the country, and not simply certain leaders.[32]

It should be noted that Dewey was not the only foreigner enmeshed with the Turkish modernization discourse and getting indigenized. He traveled with other members of a traveling library, including German Georg Kerschensteiner and Alfred Kuhne, Belgian Omer Buyse, and Swiss Albert Malche. All but Kerschensteiner were both literately and literally present in Turkey during the early years of the Republic.

"An Educational Mission"?: Dewey's Visit to Turkey and Reports for the Ministry

A few days after the law of the unification of education was passed in March 1924, Dewey was invited to Turkey. The law meant that all educational institutions, including foreign schools, were to be under the control of the Ministry of Education. Charles R. Crane,[33] a close friend of Dewey, played an important role in Dewey's acceptance of the invitation.[34] Coming to Turkey in the summer of the same year, Dewey was to have a touristic experience during his two-month stay, to act the role of diplomat negotiating the status of foreign schools, and also to work on the recommendation reports the Ministry has asked for.

Dewey arrived in Istanbul in mid-July. His first meeting was with the president of *Darülfünun* (today's Istanbul University), İsmail Hakkı Baltacıoğlu, in order to get general information about the educational system. After a week of sightseeing, Dewey started his school trips and visited about ten schools in Istanbul. Since Dewey's visit was during the summer break, he did not have the chance to see classroom practices. His observations were to be limited to school buildings, classrooms, and course materials. About a month later, Dewey moved to Ankara in order to meet the Minister of Education. He attended the Teachers Union Meeting at which he also met Atatürk. After spending about ten days in Ankara and visiting two or three schools, Dewey left for Bursa, the old capital of Ottoman Empire and traveled in the countryside. In the beginning of September, he was back in Istanbul, and about two weeks later, he left for the United States.[35]

Dewey prepared two reports for the Ministry. His first report covered issues concerning how the national budget for education should be appropriated. After returning to the United States, he sent his second report which was relatively more comprehensive in comparison with the first one. Consisting of thirty pages, it included a number of proposals, ranging from the aims of national education to the recruitment of teachers.[36] Later, he published five articles on Turkey, one of which was devoted particularly to the issue of foreign schools.[37] These articles reflected Dewey's impressions of Turkey. For example, he made an analogy between the frontier of America and the people of Turkey in terms of their openness to the change. He often reiterated the "ignorance," "backwardness," "fatalism," and "docility" of the peasants as a "fait accompli," leading also to new rules and regulations to be taken without resentment. He blessed the work of the Turkish authorities, aiming to replace "dogmatic religious inculcations" with secular nationalistic ideals and to create a "free," "independent," and "modernized" Turkey. He also pointed out that although "its loyalties are at least less dreadful than those of dogmatic religious differences," "nationalism has its evils" as well. He stressed the need for finding out "terms upon which the populations could live peaceably together" under a unified state. The closure of foreign schools, according to Dewey, was a symptom of the evil in nationalism to which the Turks have been suddenly converted and from which they would suffer most. He wrote that because of "the heritage of ignorance" and "the lack of skills, knowledge, and source for economic development," Turkey cannot afford to lose foreign assistance, if it wants to make the reforms permanent and efficient. To our knowledge, none of these articles were translated into Turkish.

In fact, very few of Dewey's works were translated into Turkish. He published numerous books and articles during his lifetime. These works dealt with many topics ranging from philosophy and ethics to education and the prospects of democratic society. Except for "*Freedom and Culture*," the nine works by Dewey that were translated into Turkish had to do with the theme of education. The same indifference to the other texts is still prevalent in the Turkish intellectual circles. This actual translation alone tells a lot about the existing discourse in this particular "life-world" into which Dewey landed and the cultural purposes he served. Through the process of selective translation, the educational ideas of Dewey were extracted from the conceptual and philosophical framework tying together his colossal number of texts dealing with a variety of themes (including education) and developed within a particular social, political, and cultural context in/to which Dewey produced his works. Through this extraction, Dewey became the figure to whom participants in the modernization project that aimed to turn the whole society into a school "pointed." It should be noted that we do not mean that the Turkish elite simply linked their discourse to Dewey as an "international authority" by distorting him. As Foucault put it, "the 'author-function' . . . is not formed spontaneously through the simple attribution of a discourse to an individual. It results from a complex operation whose purpose is to construct the rational entity we call an author."[38] In other words, it was this distortion that made Dewey's traveling into Turkey possible and turned him into a "speaker of the universal."

Reform Movements and the Reinscriptions of Dewey: Turkish Hearths, the People's Houses, and Village Institutes

During the first two decades of the Republic, there were different educational reforms that posed distinct and different problems for the intellectual elites who wanted to create a modern mode of living for the nation's population. In this section, we discuss three of these reforms. The first was the Turkish Hearths, an adult education institution that preceded the formation of the new Republic. The second was the People's Houses, another adult education initiative that was to forge the Kemalist idea of the Turkish nation. The third was the Village Institutes that trained peasant youth to become teachers. This last project was not to forge the nation through the making of the individual but instead to place the individual in a privileged position in the construction of society. By following these different reforms, it is possible to consider how Dewey enters into educational projects in relation to the broader cultural and political agendas of the Turkish Kemalist elites.

Turkish Hearths

The Turkish Hearths (*Türk Ocakları*) were established in 1908, before Dewey's visit, as well as before the establishment of the Republic, but provided part of the context that made future reforms and understandings possible. This organization had many branches around the Empire, aiming to disseminate the ideals of Turkism. After the establishment of the Republic, the Hearths were given a semiofficial status, and they expanded all around the new territory of Turkey with the mission of spreading the idea of Turkish nation. The Hearths targeted especially young people. In them, meetings were organized, classes (like instruction in the new alphabet) were taught, and libraries were established. Atatürk followed the works of these organizations closely, attended the meetings, and often praised them.[39] Having their own journals, the Hearths also provided a space for the state official and intellectual to refine the idea of Turkish nation further. The Turkish Thesis on History and the Theory of Sun Language were developed during the era in which the Hearths had operated. It was postulated that the people of the new territory were always Turks. All other cultures and languages, according to these theses, had descended from Turkish culture and language. Skirting along the borders of racism, these theses were ignoring and, in fact, working to destroy the other existing languages and cultural practices within the new territory of Turkey.[40] The same theses, at the same time, were fashioning an ambiguous national identity without its "Other" by relating all other cultures and languages to "Turks" as their originator.[41]

The 1930s were to witness that the masses of Anatolia especially had still a long way to go to join the ideals of Republican elite. The successive experiments with multiparty politics, for example, proved to be futile, leading to the unintended demonstrations against the reforms passed under the leadership of the Republican People's Party (RPP). The state elite concluded that "the people were not ready" for multiparty democracy.

The failure of the Republican elite's ideals in reaching the masses was largely attributed to the Turkish Hearths because the ideals of the Young Turks period were still dominant in these organizations.[42] The distinction between culture and civilization was still being invoked by the intellectual circles of Hearths. The leadership of the Hearths was talking of an "authentic" Turkish culture, representing a set of unique but constantly modified sentiments and attitudes.[43] They were in agreement with the ideals of the Kemalists but they were reluctant to accept the discontinuity in the Kemalist version of history that rejects the Ottoman-Islamic past. Part of the existing living cultural system, Islam, they believed, would have acted as a further unifying factor. Accordingly, their version of history narrated that Turks had met Islam at some point of their long historical path and accepted it as their religion. They placed Turks at the center of Islamic civilization. With the establishment of the Republic, the Turks were moving from

Islamic civilization to that of contemporary Western civilization. The president of the Hearths, for example, stated that their mission was to "work for the Turkish nation passing from one civilization to another."[44]

The People's Houses

In 1932, the Hearths were replaced with the People's Houses (*Halk Evleri*). Controlled directly by the RPP, the Houses were to forge the idea of the Turkish nation that was in line with the ideals of the Kemalism. Western civilization was not merely the contemporary civilization. It was the only civilization. The nation had chosen being civilized over being uncivilized, rather than choosing the contemporary civilization over the old one. More importantly, the frustrating experiment with multiparty politics proved that an idea of "authentic" culture that is apart from the state and its sacred cause had led people to mobilize in an undesirable direction. If the new regime was to survive, it was necessary to produce and spread new sentiments and attitudes to the masses.

What the Turkish nation meant in the intellectual circles went through a modification during the 1930s. More than a latent and particular pattern of beliefs and values, the idea of Turkish nation was to represent a community that lives in a given territory, shares a sense of common goals, and works for these goals with solidarity. The People's Houses were to create this national culture by amassing people into a community with a shared desire to progress and to make this desire remain permanent.[45] Dewey's ideas were integral to the founding principles of the People's Houses. As the name The People's Houses implied, the idea of Turkish nation was to be community, of which the different segments come together and work collaboratively to create a modern, developed, and democratic "We." The only division within the Turkish Society was between the "enlightened" and "the People." The Houses operated like a place for "the enlightened" to meet "the People." Seen as possessing certain qualifications, such as faith in the Kemalist ideals, the enlightened were encouraged to visit the villages in order to persuade the masses of the Anatolia of the necessity of accepting the new regulations for their salvation. The Houses and their small versions in the village (the People's Rooms) were also to deal with the welfare of "the People." The Houses and the Rooms, for example, organized lectures on how to increase agricultural productivity, how to establish producer cooperatives, and the like.[46]

At this juncture, the limited scope of the institutions of school became a concern for the state elite.[47] The existing schools were also perceived as inadequate in teaching the new collective culture. As a response to mounting problems, İsmet İnönü, the leader of the RPP, wanted a report on the conditions of primary education. İsmail Hakkı Tonguç was assigned to prepare a general report. Tonguç had studied at the graduate level in Germany. He had read Dewey, and he was particularly interested in Kerschensteiner. Until his *Köyde Eğitim* (Education in the Village) in 1938, he had published a number of works that were not directly related to peasant education but dealt with vocational and technical education. In his later works, Tonguç brought together Dewey and Kerschensteiner into a conversation that he used to outline a view of the education of peasants. In his report for the Ministry, Tonguç proposed the establishment of a new kind of school to be called Village Institutes (*Köy Enstitüleri*). After a three-year experiment, the institutes were officially founded in 1940.

The institutes operated only for six years in their original design. The Village Institutes were a project to redesign the individual who was modern. That gave them different purposes from the previous reform projects. The institutes worked against the expectations of the state struggling to be the only reference point through which the individuals define their "selves" and their world. Following Tonguç's forced resignation from the post of general director of Primary Education in 1946 and the closure of the Higher Village Institute in 1947, the drastic changes in the organization and curriculum of the institutes were introduced. After the transition into multiparty system in 1950, they were completely closed down in 1954, under the Democrat Party (DP) leadership.

Village Institutes

The broad objective of the Village Institutes was to train the peasant youth as teachers. Designed as free, post-primary, and coeducational boarding schools, the institutes were to accept only peasant children. Following their five-year education in these schools, students were to return to their own villages as teachers. Officially founded in 1940 with the passing of a law, the number of the institutes reached twenty, including one higher village institution in 1946. The same year, they enrolled more than 20,000 students, and more than 5,000 graduated.

The act, *Köy Okulları ve Enstitüleri Teşkilat Kanunu*, written by Tonguç himself also included a number of regulations concerning village schools.[48] For example, by introducing certain penalties, the law was to announce the determination of the state in making primary education compulsory. This law also made it obligatory for every villager to work at school construction and repair for a certain amount of day every year. In other words, while the future teacher of the village was being trained in the institute, the villager was to prepare the school. Since the school was to be one of the assets in the village, dedicated to its betterment and happiness, it, according to Tonguç, was the paramount duty of the villager to work for its construction and repair. In the form of *imece*, the communal cooperation and work, Tonguç went on to write in this act, were already part of the village life. The obligation to work for the village school was drawing on this tradition of *imece*. The school, built through the collective efforts of villagers, Tonguç wrote, was to turn the cooperative work of the villagers into "a conscious activity."

Tonguç explained what he meant by "the peasant problem" and "the education of the peasant" in the following way.

In opposition to what some people tend to think, the peasant issue is not an issue of development in a mechanical sense, but an issue of the reanimation of the village meaningfully and consciously from inside. The peasant should be activated and brought to consciousness as much as possible so that nobody abuses her/him in service of his own interest. No power treats the peasant as if he/she is a slave or serf. The peasant never becomes a service animal working unconscious and for free. This is what is meant by the peasant problem and also includes its educational problems.[49]

Dewey was present in Tonguç's idea that individuals are agents of their own destinies. The notion of "reanimation of the village consciously and meaningfully from inside" reflected Tonguç's insistence on the idea that the social change cannot be realized through the imposition of a preestablished end on the social body. Rather, he believed that social change occurs within, and develops out of, the social body in which individuals participate in the establishment of the ends. To say or to do otherwise, according to him, was to deny the human agency. As we go into further discussion in the following pages, the school, in Tonguç's thinking, was to provide an environment that would enable human agency and let the individuals recognize their agency, leading change and betterment in their immediate environment.

Within this rationale, Tonguç never spoke of a "peasant mentality" to be eradicated through the school. In his writings, "peasanthood" did not represent a set of preferences and dispositions that were specific to a certain group and contradicted with the idea of "progress." Tonguç viewed "village" and "peasanthood" as data revealing the realities of the environment. Peasanthood, for example, was a life "in and together with nature." The school was to be related to and was to deal with village life. Entering the school for peasant children should have meant entering an environment that reflected the realities and necessities of their village life. The school, according to Tonguç, was to provide its students with an experience through which they realized themselves as actors of change toward making the existing condition of life better and more humane for themselves and for all. Dewey's criticism of the conceptualization of education as "preparation for life" was inscribed in Tonguç's vision of school. For example, he wrote:

> If education and social life are disconnected, namely if education is not life itself, but is designed as a preparation for life on the word of some pedagogues, one day education could be rendered inept of grasping its own existential condition . . . . Life itself, means first and foremost, work.[50]

The most distinctive feature of Tonguç's writings on schooling (and of the village institutes experience in the Turkish educational history) was, perhaps, that "the present" was not cut off through/within the narratives of progress. In Tonguç's vision of social change and schooling, there was no vision of "the future" as a "thing" that comes before, and guides, change. Change, according to him, occurred constantly through purposeful activity in order to accomplish the aims, which are an outgrowth of the present condition of life. As the individuals observed and attended the very immediate of their conditions of life and worked in cooperation for the achievable aims, change became possible. He insisted that only through their involvement in the successful collaborative activity in/through which the individuals see themselves as crucial actors, a sense of self-reliance and togetherness develops in the individual, leading to the establishment of new aims and further successful actions. To have a life in a human sense in Tonguç's thinking implied consciousness of the immediate and working with others cooperatively to get control over, and better, present life conditions. The existing school, on the other hand, he argued, "has been able to educate neither their *müdavims*[51] nor the inhabitants in their immediate surroundings to have a handle on the simplest forms of occurrences among what we call natural phenomena. On the other hand, the institutions called hearths of labor and profession partially achieved this end since early periods."[52] What lied under Tonguç ideas of "work," "work activity," and work relations" was a strong criticism of the schools of the Republic. He wrote:

> Even in the ages when the care of the young generation was instinctive, the primary interest has been to educate them for work by engaging them actively in a life of work. This is an obvious fact in the cultural history of humanity. On a spectrum ranging from the most primitive type of work life to the highest level of culture, the picture remains the same: all pieces of work were created through the combination of manual work, intellectual activity and thinking. Human beings have been educated by partaking in the process of creating these works. However, in many societies, the schools have still not gone beyond educating only a portion of their members, namely the privileged groups. Education given in the family and the workplaces is unlike this. In contrast to schools, these settings are not concerned with creating social layers by excessively developing the abilities of certain individuals in an artificial manner. This huge disparity between the educational institutions of school and family/work place stems from the fact that schools are unaware of the educational role and value of work, and still have not been able to embrace all members of the society. . . . One of the most important duties of our century and the ones to come will be without doubt to eliminate this discrepancy.[53]

Tonguç wrote against the widespread tendency to assume that in order to bring people under a set of ideals and mobilize them for a better future, they had to first be taught a set of principles, values, knowledge, and dispositions. By reiterating "the relation among work activity, intellectual activity and thinking," Tonguç rejected the idea that all these principles, values, knowledge, and ideals existed outside and before the actions and thinking of the individual. "The school," he wrote, "lost its connection, while it is turned into an institution that deals with every single problem" and "its bookish character falls short in being rational and did not overlap with the realities of the life."[54] Furthermore, the attribution of a special status to knowledge, according to Tonguç, led to the fragmentation of society, creating social layers in accordance to who posses it and who does not. Dewey was present in Tonguç's account of knowledge as a means. According to Tonguç, the value of knowledge came from the extent to which it leads to successful action. As long as it attains relevance in life and makes the individual capable of identifying and coping with the immediate problems effectively and successfully, according to him, knowledge had a value. He wrote:

> In the village institutes and the village schools, knowledge is an instrument. This instrument will be used to stimulate students' intellects, to incite their imagination, and to teach them basic work/study methods. Putting stress on the acquisition of simple information will never be compatible with the spirit of work so fundamental to the village schools and the village institutes. The village institutes will bring up simple yet skilled (goal-getter) individuals who will be citizens deriving spiritual strength from their abilities and accomplishing tasks with their intellectual power.[55]

Following the rationale discussed here, twenty-one institutes were established across Turkey. Aiming to accommodate themselves to local conditions and needs, each institute covered three or four provinces only. five percent of the institutes' curriculum was devoted to "culture classes." These classes included Turkish, history, geography, citizenship, mathematic, chemistry, foreign language, physics, drawing, physical education, folk dances, music, home economics and child care, pedagogy, cooperatives, and agricultural economics. The remaining 50 percent was divided into two. Twenty percent of the curriculum was reserved for agriculture classes. These classes were field crops, horticulture, industrial crops breeding, animal science, poultry production, honey bee production, aquaculture, and fishing. The remaining 25 percent consisted of technical classes, including blacksmithing, carpentry, building construction, and handicrafts. Each institute had autonomy in defining their weekly, monthly, yearly programs in accordance to their needs.[56]

The institutes were designed as work schools. All cultural subjects were taught within the context of and in connection with the work activities around which the subject revolved. Each institute had lands and its own workshop. Organized as a work community, in the institutes the students work cooperatively to cultivate crops, to construct institute buildings and to raise animals. Using knowledge to perfect the work skills needed in village life, the products were marketed and used the profit to meet the other needs of the institute. The students were involved in every step of work activities, from the establishment of goals to evaluation. The task of the teacher was to assist the students with her advice. "Saturday meetings" were an important part of the institute life. Every Saturday, the members of the institute gathered to evaluate the week and solve the problems and crisis.

The village institutes embodied a notion of liberal, individualistic society that worked against earlier Kemalist visions of the solidarity and coherence of the nation. As discussed earlier, the village institutes were a project that placed the individual in a privileged position in the construction of the nation. They did not seek to create the nation through making the individual. The distinctive feature of the Institutes Project was that it sought to connect individuals through their relationship to each other, not through the state. This distinctive feature of the village institution was not really discussed by Turkish academic circles. Instead, today there is a resurgence of nostalgia for the Village Institutes.

Concluding Thoughts

In this chapter, we have argued against the widespread tendency in the academic circles to construing the present disjunctures in the educational arenas of countries across the world as incompleteness, deviation, and so on. Drawing on the recent poststructural and postcolonial scholarship, we rejected the examination of the encounter of the "global" with the "particular" within the problematic of "impact," "influence," and/or "importation." Through our approach toward Dewey as "indigenous foreigner," we argued that ideas, discourses, and texts are not fixed entities, and they do not drop on a tabula rasa as they circulate across the globe. Rather, we insisted they are always altered, manipulated and distorted by the culture into which they penetrate and circulate.

We also argue that Dewey as an indigenous foreigner does not stay the same as his ideas move in different traveling libraries to embody different notions of modernity. Dewey's assemblage in the Turkish Hearths, the People's Houses, and Village Institutes, are inscribed differently in the changing educational, political, and cultural narratives as new problems are posed, and the concerns about the future change. For example, we argued that during the People's Houses Project, Dewey was a name blessing the "enlightened elite" helping the masses of Anatolia reach to the moral level of believing the ideals of the regime. When the Village Institutes were constructed, Dewey was inscribed in the narrative of the village that was reanimating from what was viewed as bringing out in the child was innate and natural to them.

The transmogrifications of Dewey as he "traveled" with other sets of ideas were "productive." By productive, we mean the cultural and social processes in which these ideas were engaged contributed to changes in ideas, institutions, and the patterns of social relations. As we noted in the beginning of this chapter, individuals were produced who made claims of rights for linguistic and cultural recognition. The notions of individuals as actors who could made claims were predicated on the constructions embodied in the pedagogical "child."

It is easy to freeze and fixate the issues of schooling and modernization in contemporary Turkey as one of permanent crisis. To take such a stance is to make their solution impossible. Our approach is to place the issues and problems of Turkey and its education as the deconstruction of the idea of modernity in Turkey. We have problematized the sacredness of revolution at the beginning of the century and its imagined coherence as a march toward contemporary civilization. By historicizing Dewey's connections and assemblages within Turkish education, our analysis has engaged in/with the tensions of modernity in Turkey. We looked at the ideas of modernity in Turkey, such as "education" as the application of a set of "universal" and "natural" principles (sold, e.g., by Dewey). That "selling" was seen as having no interest in the social, political, cultural context in which/to which Dewey produced his works. Having no concern with, and no recognition of, the social, cultural, and political dimension in his writings on education, circulated as cultural practices, the intellectual circles of Turkey continue to see Dewey merely as a pedagogue. While in many other contexts, the interest in Dewey increasingly grew and led to a massive amount of secondary literature on Dewey, this kind of literature does not exist in Turkey. The aim of this chapter is to provide a corrective to this omission in Turkish educational as well as social thought.

Notes

1. We would like to thank Thomas Popkewitz for his very crucial comments, his support, and his patience, as we worked on this chapter. We are also grateful to Marianne Bloch for giving us very helpful comments. We want to thank members of Thursday and Wednesday reading groups at UW-Madison for reading and responding to earlier drafts of this chapter, including Dar Weyenberg, Dory Lightfoot, Ruth Peach, Barb Tarockoff, and Noah Sobe. We also thank Cengiz Sürücü for his comments on the earlier version of this chapter. Finally, we thank Dory Lightfoot, Akile Zorlu, and Kaan Durukan for help with English.
2. Meltem Ahıska, "Occidentalism: The historical fantasy of the modern," *The South Atlantic Quarterly* 102, no. 3 (2003): 356.

3. Ahmet İnsel, "Eğitimde positif ayrımcılık gerekli." *Radikal*, September 9, 2004.

4. Meltem Ahıska, "Occidentalism," 367.

5. Meltem Ahıska, "Occidentalism," 356.

6. For example, see Sabri Büyükdüvenci, "John Dewey's impact on Turkish education," in *The New Scholarship on Dewey*, ed. Jim Garrison (Norwell, MA: Kluwer Academic Publishers, 1995): 393–400; Selahattin Turan, "John Dewey's report of 1924 and his recommendations on the Turkish educational system revisited," *History of Education* 29, no. 6 (2002): 455–543; Ernest Wolf-Gazo, "John Dewey in Turkey: An educational mission," *Journal of American Studies of Turkey* 3 (1996): 15–42; Huseyin Bal, *1924 raporunun Türk eğitimine etkileri ve John Dewey'in eğitim felsefesi* (İstanbul: Kor, 1991); William W. Brickman, "The Turkish cultural and educational revolution: John Dewey's Report of 1924," *Western European Education* 16, no. 4 (1984): 3–18.

7. Benedict Anderson, *Imagined Communities: Reflections on the Origin and Spread of Nationalism* (New York: Verso Press, 1991).

8. Akhil Gupta and James Ferguson, "Beyond 'culture': Space, identity, and the politics of difference," in *Culture, Power, Place: Explorations in Critical Anthropology*, ed. Akhil Gupta and James Ferguson (Durham, NC and London: Duke University Press, 1997), 33–51.

9. See Thomas S. Popkewitz, chapter 1, this volume.

10. Dipesh Chakrabarty, "Universalism and belonging in the logic of capital," in *Cosmopolitanism*, ed. Carol A. Breckenridge, Sheldon Pollock, Homi K. Bhabha, and Dipesh Chakrabarty (Durham, NC and London: Duke University Press, 2002), 106.

11. Chakrabarty, "Universalism and belonging," 106.

12. Gupta and Ferguson, "Beyond 'culture': Space."

13. Wolf-Gazo, "John Dewey in Turkey: An educational mission."

14. Bernard Yack, *The Fetishism of Modernities: Epochal Self-Consciousness in Contemporary Social and Political Thought* (Notre Dame, IN: University of Notre Dame Press, 1997).

15. Peter Wagner, *A Sociology of Modernity: Liberty and Discipline* (New York: Routledge, 1994), 20.

16. Şerif Mardin, "Religion and secularism in Turkey," in *Atatürk: Founder of a Modern State*, ed. Ali Kazancıgil and Ergun Özbudun (London: C. Hurst & Co. Ltd., 1981), 208–209.

17. Ali Kazancıgil, "The Ottoman-Turkish state and Kemalism," in *Atatürk: Founder of a Modern state*, 38.

18. Ergun Özbudun and Ali Kazancıgil, "Introduction," in *Atatürk: Founder of a Modern State*, 3.

19. İnsel and Kazancıgil use Max Weber's notion of patrimonialism to discuss the political culture of the Ottoman Empire, characterized by the unquestionable power of the state and the lack of civil society. Ahmet İnsel, *Türkiye Toplumunun Bunalımı*, 2nd ed. (İstanbul: Birikim Yayınları, 1995); Ahmet İnsel, *Düzen ve Kalkınma Kıskacında Türkiye* (İstanbul: Ayrıntı Yayınları, 1996), Ali Kazancıgil, "The Ottoman-Turkish state and Kemalism." See also

Kemal H. Karpat, "Historical continuity and identity change: How to be Modern Muslim, Ottoman and Turk," in *Ottoman Past and Today's Turkey*, ed. Kemal H. Karpat (Boston: Brill, 2000), 1–28.

20. İnsel, *Türkiye Toplumunun Bunalımı.*

21. Rulership of Islam.

22. İnsel, *Türkiye Toplumunun Bunalımı.*

23. In 1830, the grand admiral of the Empire, e.g., was to declare: "I am back from a visit to Russia. On my return, I became more than ever convinced that, if we are further delayed from imitating Europe, we shall be left with no alternative to the obligation of going back to Asia" (cited in Ali Kazancıgil, "The Ottoman-Turkish state and Kemalism," 38).

24. İnsel, *Düzen ve kalkınma kıskacında Türkiye*, Şerif Mardin, "Religion and secularism in Turkey," and Ali Kazancıgil, "The Ottoman-Turkish state and Kemalism."

25. For example, the establishment of the school of administration (*Mülkiye*) was part of this project. The secularization of higher education had started before 1839 with the founding of the Military Academy and the School of Medicine. During this era of *Tanzimat* (1839–1876), the idea of free primary education for all began to be articulated by the Ottoman state elite. In 1846, the Ministry of Public Instruction was created, and the state converted the schools of the non-Muslim subjects that were financed by private support or by charitable grants into a state-financed primary school. The 1850s and 1860s were to witness the establishment of post-primary secular state schools (*Rüstiye, İdadi, and Sultani*) in Istanbul and many provincial centers. Along with secular state schools, the traditional Islamic schools, such as *Mektep* and *Medrese* continued to operate. See Mardin, "Religion and secularism in Turkey."

26. Mardin, "Religion and secularism in Turkey," 206.

27. Mustafa Kemal Atatürk, e.g., was born in Selonika (now, Thessaloníki, Greece), attended the military middle school there, entered the Manastır Military High School, and then progressed to the Military Academy in İstanbul.

28. Mardin, "Religion and secularism in Turkey," 205.

29. Ataturk wrote, e.g.: "It is futile to try to resist the thunderous advance of civilization, for it has no pity on those are ignorant or rebellious. . . . We cannot afford to hesitate any more. We have to move forward . . . Civilization is such a fire that it burns and destroys those who ignore it. Cited in Ahıska, "Occidentalism," 379.

30. The idea of preparing people for the future democracy finds its expression in the following words of Mustafa Kemal Atatürk: "Turkey is going to build up a perfect democracy. How can there be a perfect democracy with half the country in bondage? In two years from now, every woman must be freed from this useless tyranny. Every man will wear a hat instead of a fez and every woman will have her face uncovered; woman's help is absolutely necessary and she must have full freedom in order to take her share of her country's burden." Cited in Ertan Aydin, *The peculiarities of Turkish revolutionary ideology in the 1930s: The Ulku version of Kemalism, 1933–1936* (Ph.D. dissertation, Bilkent University, Ankara, 2003).

31. Fay K. Berkes, *The Village Institute Movement of Turkey: An Educational Mobilization for Social Change* (Ph.D. Dissertation, New York, Columbia University, 1960), 42.

32. John Dewey, *The John Dewey Report* (English translation from its Second publication in 1952) (Ankara: Milli Egitim Bakanligi, Test ve Arastirma Bürosu, 1939/1960), 1.

33. Chicago businessman, philanthropist, and trustee of the American College for Girls in Istanbul, Crane was a member of President Wilson's Special Diplomatic Commission to Russia in 1917; a member of the American Section of the Paris Peace Conference; American Commissioner on Mandates in Turkey in 1919; and U.S. ambassador to China in 1920 (See Wolf-Gazo, "John Dewey in Turkey: An educational mission").

34. Brickman, "The Turkish cultural and educational revolution."

35. For detailed information on the journey of Dewey, see Bahri Ata, "1924 Türk basını ışığında Amerikalı eğitimci John Dewey'nin Türkiye seyahati," *G. Ü. Gazi Eğitim Fakültesi Dergisi* Cilt: 21, Sayı: 3 (2001): 193–207.

36. As mentioned before, as well as stressing the importance of well-stated aims of education and blessing the enlightening role of the elite in preparing individuals for participating, in this report, Dewey reminded the Ministry not to confuse unity with uniformity and to ensure that schools accommodate themselves to the varying conditions and needs of localities. He emphasized the necessity for improving the financial and social status of teachers. Dewey also wrote that "there must be distinct types of normal schools for the training of rural school teachers, with especially reference to the needs of those who toil on the soil, and are the mainstay of Turkish life." He recommended putting the efforts into translating foreign educational literature, especially that of progressive schools, in Turkish, establishing libraries, and sending the teachers abroad, at least to the foreign schools in Turkey in order to let them learn about progressive teaching (see John Dewey, *The John Dewey report*, 18–19).

37. John Dewey, "The Problem of Turkey," in John Dewey, *The Later Works, 1925–1953*. Vol. 2, ed. Jo Ann Boydston, Bridget A. Walsh, textual ed. (Carbondale and Edwardsville: Southern Illinois University Press, 1925/1984), 189–198; John Dewey, "The Turkish Traged." In *Impressions of Soviet Russia and the Revolutionary World: Mexico, China and Turkey* (New York: New Repuclic Inc, 1929), 197–207. Dewey, "Angora, the New," in *Impressions of Soviet Russia*, 208–219; Dewey, "Secularizing a Theocracy: Young Turkey and the Caliphate," in *Impressions of Societ Russia*, 220–234; John Dewey, "Foreign Schools in Turkey," *New Republic*, 41 (1925): 40–42.

38. Michel Foucault, "What is an author?" in *Language, Counter-Memory, Practice: Selected Essays and Interviews*, ed. Donald F. Bouchard, trans. Donald F. Bouchard and Sherry Simon (Ithaca, NY: Cornell University Press, 1977), 127.

39. Holly A. Shissler, *Between Two Empires: Ahmet Ağaoğlu and the New Turkey* (New York: I. B. Tauris & Co Ltd., 2003).

40. In the congresses of Turkish Hearths, e.g., it was often stressed that cultural and language diversity should not be embraced. In these meetings, fees and

even imprisonment were proposed for those who spoke languages other than Turkish. In 1928, "Citizen, Speak Turkish!" Champaign (*Vatandaş Türkçe Konuş!*) was also organized by these organizations. Similar works were sometimes taken on by their successors, the People's Houses. See Füsun Üstel, *Türk Ocakları, 1912–1931* (İstanbul: İletişim Yayınları, 1997). See also Mustafa Çapar, *Türk Ulusal Eğitim Sisteminde Öteki ve Ötekiye Yaklaşım* (Ph.D. dissertation, Hacettepe Üniversitesi, Ankara, 2004).

41. Sevan Nişanyan, "Türk Kime Denir?" *Modernleşme ve Çokkültürlülük*, ed. Nazan Aksoy and Melek Ulagay (İstanbul: İletişim Yayınları, 2001), 198–212.
42. Aydin, *The Peculiarities of Turkish Revolutionary Ideology in the 1930s*.
43. Shissler, *Between Two Empires: Ahmet Ağaoğlu and the New Turkey*.
44. Cited in Aydin, *The Peculiarities of Turkish Revolutionary Ideology in the 1930*.
45. Aydin, *The Peculiarities of Turkish Revolutionary Ideology in the 1930s*.
46. Asım Karaömerlioğlu, "The People's Houses and the cult of the peasant in Turkey," *Middle Eastern Studies* 34, no. 4 (1998), 67–91.
47. For example, according to 1935 statistics, more than 80% of the population was living in rural areas. Only 16% of the village children had a school to go to.
48. *Köy Okullari ve Enstitüleri Teskilat Kanunu Izahnamesi* (Ankara: Maarif Vekilligi Nesriyat Müdürlügü, 1943).
49. İsmail H. Tonguç, *Canlandırılacak Köy* (İkinci Bası) (Istanbul: Remzi Kitabevi, 1947), 85.
50. Cited in Pakize Türkoğlu, *Tonguç ve Enstitüleri* (İstanbul: Yapı Kredi Kültür Sanat Yayınları, 1997), 158.
51. Instead of "student," Tonguç often used the word *müdavim*, meaning a person who goes to a place regularly and implying no power hierarchy embedded in it.
52. İsmail H. Tonguç, *Köyde Eğitim* (İstanbul: Kültür Bakanlığı İlkokul Öğretmen Klavuzları, 1938), 165.
53. Tonguç, *Köyde Eğitim*, 166.
54. Tonguç, *Köyde Eğitim*, 178
55. *Köy Okullari ve Enstitüleri Teskilat Kanunu Izahnamesi*, 111.
56. Yahya Özsoy, *Köy Enstitülerinde Öğretim Programları* (Ankara: Köy Enstitüleri ve Çağdaş Eğitim Vakfı Yayınları, 1997).

IV

The Americas

8

Discursive Inscriptions in the Fabrication of a Modern Self: Mexican Educational Appropriations of Dewey's Writings

· *Rosa N. Buenfil Burgos*

Dewey's writings, just as any other text, in their articulation with other discourses available at a certain time and space, are resignified in such a way that they can become just the opposite of what they previously were.

Dewey as an *indigenous foreigner*, is an interesting example of historization that puts in operation images of intelligibility that challenge traditional logics and common sense. Paradox and aporia are the two logics that come to my mind with such an expression. In addition, the *indigenous foreigner* also brings to the fore the ceaseless circulation of ideas as a globalizing feature of the beginning of the twentieth century. Dewey's prospects have been successively articulated in surfaces of "global" and local discourses, which is the case in the Mexican inscriptions that will be commented here. As ideas circulate, they get resignified and inscribe marks into the modernizing of the nation and the individual. These patterns may be intraceable one to one to the original source, however, their remainders can be found in different discursive surfaces, and this is precisely the aim of this chapter.

My chapter does not deal with Dewey's pragmatism as such, but with the processes whereby it was inscribed in Mexico under specific circumstances.[1] I try to elaborate two different appropriations of Dewey's work that have taken place in Mexico. The first can be located in the 1920s and is closely related to a revolutionary atmosphere; the second reading can be placed in the 1970s and is embedded in the increasing visibility of Marxism in Latin America. My focus on these two moments does not amount to ignore other possible appropriations. Just to mention one important inscription of Dewey's prospect in Mexico, I can evoke that in the late 1930s, when Trotsky came to live in Mexico running away from Stalinism, and was hosted by the nonorthodox left-wing intellectuals, John Dewey visited him in Mexico.[2] Dewey gathered with Mexican socialist intellectuals and activists in several international antiwar meetings in which he had a salient role. With my decision to deal only with two inscriptions, I try to produce an image of Dewey as an indigenous foreigner based on two theses: His educational ideas are altered in the very process of articulation with local discourses, and in this process they are occupied and colonized thus fabricating a Dewey who is more and less than John Dewey.

Given the relational character of any discourse, no analysis is possible lacking the consideration of other discursive series contiguous to the one under scrutiny.[3] The normalized image to evoke this is by means of the idea of context. This notion however, must not be regarded as a mere background or environment, but as the very symbolic structuring that enables a discourse to be such. Context in this chapter, therefore will be understood as a condition of possibility for the appropriation and thus *iteration*[4] of Dewey's pedagogical thought. These conditions involve series different in kind such as the intellectual atmosphere, geopolitical conditions, religious, cultural, and economic settings, and whatever tendency may be articulating symbolic life at a certain moment. Accordingly, I present these conditions in each specific inscription to provide the ground for a plausible interpretation.

First Inscription, Dewey in Mexico in the 1920s

Some features can be stressed as conditions of possibility for Dewey's educational ideas in Mexico.

First, the intellectual atmosphere of the second half of the nineteenth century involved tensions between on the one hand, Enlightenment, positivism, and political liberalism; and on the other hand, a deeply rooted Roman Catholicism. These tensions took place within the liberal government implemented by Benito Juarez who contended against the political, economic, and intellectual empire the Catholic Church held from the Spanish conquest onward. The reform that took place from 1856 onward precisely excluded the rule of Roman Catholicism. The reform separated the Church from the formal institutions of the Mexican political system, and removed its predominance over public matters such as education and the legal system among other realms. This process was accompanied by the rise of a positivistic intellectual atmosphere that prevailed for many years, with the discursive scaffolding of values, programs, and institutions such as the Escuela Preparatoria,[5] which was created around the positivistic values of progress and modernity. This articulation, however, did not manage to extirpate the religious imaginary of salvation from the Mexican population. The Catholic Church, on its part, has never accepted this exclusion, and has fought back on different fronts; legal, religious, moral, economic, and political strategies have been displayed in their effort to maintain prevalence over laic political and public institutions.

The Mexican revolution (from 1910 to the first postrevolutionary phase of stability in 1917), rose against the thirty-year dictatorship of a liberal government (of Porfirio Díaz) that enhanced Positivism in intellectual terms and articulated it with a religious approach in moral matters. The Mexican revolution was articulated around an *overdetermined*[6] agenda composed by heterogeneous intellectual and political sources: liberalism, socialism, rationalism, positivism, Enlightenment, political and social demands (posed by peasants, workers, professionals, and other social sectors), amongst the most salient sources. This revolution can be represented as a political, social, and cultural dislocation, ripping the social fabric into threads,[7] and setting the conditions for the need of a salvation narrative (i.e., the Mexican Revolutionary *Mystique*).[8]

Moreover, the intellectual and political composition that emerged from this unity of diversity was characterized by hybridity and eclecticism, which permeated the national revolutionary program. It involved social demands as expressed by its popular bases (e.g., land, justice, and so on), the political conceptions informing specific plans of each social sector, intellectual traditions differently rooted in geographical areas, and last but not least, the inescapable common sense dwelling in Mexican idiosyncrasy. The latter, in turn, combined religious fanaticism, as well as its anticlerical counterpart, it coupled national chauvinism and *malinchismo*, it united naive optimism and sour pessimism.[9]

Second, the educational crusade in the context of the Mexican revolution consisted of an Enlightenment optimist agenda revisited and amalgamated with revolutionary hopes. In 1921 the Ministry of

Education was restored after the armed movement and charged with producing a new modern Mexican.[10] J. Vasconcelos was in command and he proposed an ambitious program to restore the nation by means of a vigorous educational strategy. It involved the creation of an enterprising rural schooling scheme, the recruitment of an army of school teachers, the creation of itinerant teams to train teachers and educate communities (called *Misiones culturales* trans. Cultural missions). Vasconcelos brought together enterprising and creative educators, set up offices, institutions, and dependencies, and built libraries in rural sites, among other strategies. All this vitality produced an adequate setting for the introduction of Dewey's democratic and socializing principles. It must be stressed, however, that Vasconcelos in his confrontation with the previous intellectual regime (i.e., Positivism) claimed to adhere a "spiritualist" philosophical approach to education, thus excluding the excessive faith in reason, science, and empiricism that pervaded the prerevolutionary educational programs.

A third condition of possibility of Dewey's inscription, is related with the contacts that took place between Dewey and Mexico by scholars and students. There were three scholars who had direct contact with John Dewey and his educational theories: Moisés Sáenz, Manuel Gamio, and Rafael Ramírez.[11] During the postrevolutionary years, the nation had to be reconstructed and Dewey's pedagogical thinking was considered to be adequate to these ends. Accordingly, Mexicans had traveled to the United States to learn from Dewey. Whatever these scholars were able to understand from his thought was introduced to organize the Mexican revolutionary school and to contribute in the formation of the nation. Sáenz and Ramírez who occupied key positions in the Ministry of Education during this decade, were convinced that Dewey's progressive schemes fitted nicely with the political and intellectual hopes of the Mexican revolutionary education. Historians of both the Mexican revolution and the educational system agree on the point that Dewey's precepts achieved their best implementation in rural education,[12] however, the whole educational scheme was permeated by his principles.

Fourth, the Mexican revolutionary mystique (MRM onward) was produced as a promise of plenitude or a salvation narrative. This hybrid system of signification of the Mexican revolution emerged and operated as a mystical discourse. It came to provide a strong feeling of belonging and transcendental union; it gave meaning to the sacrifice and the loss, offering a horizon of compensation and an idyllic image of the future. It provided new images of identification for indigenous groups, peasants, industrial workers, and schoolteachers, as well as the "champions of the revolution," whose dignity had to be restored to its correct position. It cemented the dispersed agents and ideals involved in the armed and political movements, and operated as the ultimate source of legitimacy and proof of the righteousness of policies, laws, rituals, institutions, budgets, and so forth. Considering the constitutive fanaticism of the population it is not surprising that the symbols of this mystical discourse came to occupy the sacred positions formerly devoted to religious emblems. It is neither surprising that the new ruling bloc later became aware of this and has strategically benefited from it. The MRM was inhabited by legendary heroes and epic events[13] being later immortalized in literature,[14] murals,[15] popular and classic music, in agrarian and school rituals, in unionist agendas and later, in the film industry.

Finally, Dewey visited Mexico to supervise the educational laboratory: the School of Action. By 1926 he was invited to Mexico to observe and assess the way in which his ideas and precepts had been actualized in Mexican schools. He gave conferences in the Faculty of Philosophy, the Summer School, and the Society of Language Teachers. This was his most important visit, and it took place in July. His visit involved conferences in both the Ministry of Education and the National University of Mexico, and visits to different school sites in the countryside: Xocoyuca, Panotla in Veracruz, the Misión Cultural in Santa Cruz, and a rural school in Tlaxcala.[16] It was said that Mexico was the first country in which Dewey's ideas had been put into practice in the Programa de Acción.

These and many other are conditions in which a Mexican discourse articulates Dewey's intellectual dispositive with the horizon of plenitude produced in and by the Mexican revolution.

This first inscription involves an assemblage/articulation of Dewey's ideas and the MRM in the educational crusade, in the 1920s. I argue that the revolution produces a social and cultural dislocation, which is a condition for the need of a salvation narrative (i.e., the MRM). Tracing the displacements and the flow or circulation of ideas it is possible to study the construction of equivalencies between Dewey's educational program and the Mexican educational crusade (i.e., cultural convergence and contact). This is an articulation of, on the one hand, Enlightenment, rationalism, pragmatism, and the New School;[17] and on the other hand, Mexican liberalism, democracy, socialism, and republicanism thus producing the impure/overdetermined character of the revolutionary discourse. Education is amongst the five priorities of the revolutionary program and within it the School of Action is constructed as a key reference of the educational reform.

The previously mentioned conditions set the symbolic structures under which Dewey's proposals are introduced and iterated. The hybridity and overdetermination of the MRM, and the confronted imaginaries competing in those days, permeated the articulations of his ideas with the educational crusade. It is possible to find all sorts of inscriptions producing different combinations of the native and the foreigner. Allow me to give some examples.

Vasconcelos, very much against empiricism, in his *Antología de textos sobre educación*[18] takes up against Dewey (and his Mexican followers) in a rather invalidating manner. First of all, Dewey is portrayed as the incarnation of Positivism which was in turn constructed as the universal representative of artlessness and lack of high intellectual values. Then Vasconcelos takes *all learning by doing* as a motto around which he criticizes concepts and principles on multiple occasions, however, not from different angles. Vasconcelos especially disapproves of the following concepts:

- mind as an adaptable (adjustable) faculty whose function would be to guide intelligent action in a changing world,
- learning through experience and personal interest,
- human beings as social beings, and
- the classroom as a small society, the relationship between school and society.

According to Vasconcelos's smetaphysical essentialism, Dewey's evolutionism would not understand the difference between animals and human beings, and in giving too much attention to minute things (such as "to properly wear a tie and knot his shoelaces") he would neglect huge transcendental issues. He also criticizes both "Rousseau and Dewey for teaching lessons already taught three thousand years before by the Indian Vedanta and later retrieved by the Socratics."[19] Other motives for his rejection are that Dewey's pragmatism and schemes address the European immigrant children living around factories in shanty towns in the United States of America, and this social reality has nothing to do whatsoever with the Mexican child condition This is the flavoring of his resignification of Dewey's thinking.

Strongly contrasting with the previous, Rafael Ramírez, a pioneer in the Misiones Culturales, published in 1924 *La escuela de la acción en la educación rural* (trans. The School of Action in Rural Education) and *Escuela técnica para campesinos* (trans. Technical School for Peasants),[20] to put in extremely simple words the pragmatic aspects of a school that would start from zero and fabricate the new modern Mexican. In its pages, there is no single reference to any theoretical source, nonetheless, no sagacity is needed to find traces of Dewey's pragmatism and democratic values there. In the early 1930s however, Ramírez's two main conceptual sources start to be visible. In 1935, he published *La escuela proletaria: Cuatro conferencias sobre educación socialista* (trans. The Proletarian School: Four Conferences about Socialist Education),[21] in which a eulogy of U.S. progressive school is the subject of the second conference. Dewey's presence is more visible in the book *Técnica de la enseñanza* (trans. Teaching Technique)[22] published in 1937 to provide a textbook for the students of *Escuelas Regionales Campesinas* (trans. Vernacular Peasant Schools).[23] These schools were established to train the rural schoolteachers. Ramírez praises Dewey's paper *Interest in Relation to Training of the Will*[24] and also *A Pedagogical Experiment*[25] drawing from both teaching notions and recommendations such as: the concept of experience, concrete activities in school as the outset for education, and the child interest as the departing point for school activities; school activities emulating vital activities, observation, sensitive and cognitive action instead of verbal and mental acts, motivation and encouragement as a reinforcement of child interest, inter alia. Learning is understood as something that cannot be received from someone, but something resulting from the learner's experience.

Moisés Sáenz promoted indigenous education in Mexico and was strongly convinced that Dewey's pedagogical contributions were precisely what the Mexican revolutionary school needed if it was going to unify the nation, socialize the children, and promote reasonable forms of life, especially, for rural communities.[26] In 1926 he gave a lecture in the University of Chicago, sponsored by the Harris Foundation. The lecture was about Mexican integration through education. Sáenz presented a pragmatic approach to education that had explicit references to Dewey's philosophy and pedagogy. Insistently quoting *School and Society*,[27] he evoked key concepts such as motivation, respect for the child personality, self-expression, vitality of school work, the method of projects, learning by doing, and democracy through education.[28] In less academic interventions perhaps Dewey's name would not be written or pronounced, however, his pragmatism and educational viewpoints would permeate Sáenz's schemes and programs for rural, and especially, indigenous schooling, and as a means to bring modern techniques to produce the new Mexican.

No matter how distant could Sáenz and Gamio be in anthropological debates (e.g., "integration versus incorporation" of the indigenous into the nation), there was always an agreement concerning their pragmatic approach on indigenous educational programs.

So far, some key leaders of education during the 1920s have been paraphrased to provide an idea of the way in which Dewey's thought penetrated the postrevolutionary educational program. This involves a fusion with other available educational and political discourses, with the social basic claims for land, peace, justice, and so on, as they were understood by intellectuals, leaders, translators, interpreters, and other mediators. While being part of other social domains (e.g., agrarian, labor, ethnic, and so on) these demands circulated throughout each of these spheres, and in this traveling they partially altered each other, thus producing for instance, the petition for peasant or indigenous education. To give just an example of the intellectual fusion within the educational domain, let me mention that Ramírez and Sáenz both studied at the Normal de Xalapa in Veracruz, where the great pedagogical influence of the Spanish Rationalist Ferrer Guardia had been spread. This tradition disseminated in the south of Mexico (Campeche, Yucatán, Tabasco, and Veracruz) where it was later articulated with the socialist experiment promoted by Carrillo Puerto and Alvarado (in Yucatán in the early 1920s). In addition Rebsamen, head of the Normal de Xalapa, lectured on Positivism (e.g., Comte and Spencer). Thus it is easy to observe the intense overdetermination (i.e., circulation and fusion) of educational traditions that were articulated with Dewey's thought.

A constitutive ambiguity must be stressed here. Deweyean Pragmatism in the United States was associated with progressive ideas (e.g., democracy, antiwar groupings, antiauthoritarianism) some of which had been also taken up by the left wing.[29] But this association was not shared by the Mexicans since, on the one hand, the Left was represented by Socialism and even Communism. On the other hand, philosophical Pragmatism was linked to Positivism which was, in turn, seen as part of the political position promoted by the prerevolutionary ruling bloc (the liberal dictator Porfirio Diaz).

Thus, some intellectual and political contorsions had to take place in order to accommodate a pragmatist who at home was in the progressive side. In Mexico his philosophical ideas were associated with the prerevolutionary enemy (just remember Vasconcelos's bitter rejection of Deweyean principles) and, could, however, fit the revolutionary discourse by means of a series of equivalencies produced by some educational leaders who met Dewey in his environment (i.e., as a progressive intellectual) and managed to introduce him as part of the revolutionary narrative.

This first educational reading of Dewey's writings is constitutive of the way in which Mexican official revolutionary culture, bringing together Enlightenment-reason-science with democratic, socialist, liberal, and populist hopes of progress, produces the educational experiment: Dewey *á la Mexicana* (in the Mexican style) which makes visible one aspect of the regulatory discourse toward the fabrication of another modern self in the midst of tensions between individual and the community.

Second Scenario, Dewey in the 1970s

Some conditions of possibility can be distinguished that allow a different appropriation of Dewey's work in Mexico. This second discussion is a less enthusiastic and more critical scrutiny of both philosophical and educational pragmatism. The 1970s in Mexico, as in many other countries in the world, had witnessed too many international political mistakes (the two world wars, the Holocaust, Stalinism, invasions, and so on), and the hopes of progressive intellectuals, especially in Latin American countries, were leaning on socialist ideals. Let us move to these conditions.

First are the conditions of the political culture of Mexico, a nation that had been ruled by a one-party political system since 1929. From the years in which the military phase of Mexican revolution finished (i.e., the national movement around the 1920s, and the local and ephemeral uprisings in the 1930s), and as a sign of political advance, in 1929 the group that managed to stabilize the country founded the Partido Nacional Revolucionario (PNR for its initials in Spanish). It was reformed in 1938 during the so-called Socialist regime led by Lázaro Cárdenas, and then became the Partido de la Revolución Mexicana (PRM for its initials in Spanish). In 1939 the world was shaken due to World War II, conservative Mexicans had been panicking because of the so-called socialist regime introduced by Cárdenas (1934–1940). This conservative wing formed a strong front that was paradoxically inclined to liberalism in economic management, religion, and U.S. values in cultural terms, and against Socialism in political terms. They also formed a party, the Partido Acción Nacional, to compete against the PRM. The latter was said to have won the 1940 elections, however, an important price was paid for this: the exclusion of the socialist threat from the ruling party, the "rectification" of the Mexican revolutionary course, and a reinforced subordination to U.S. prescriptions. By 1945 the PRM became the notorious Partido Revolucionario Institucional (PRI for its initials in Spanish) that has won formal elections and ruled continuously from then to 2000 (formally fifty-five years, and actually seventy years—from 1929 to 2000). By the 1970s, the PRI had normalized and naturalized its forms and means to rule the nation. However, many revolutionary social demands had not been met. Disperse antagonistic foci had emerged creating a conflictive scenario: peasant rebellions, railroad workers' conflicts, a medical practitioner's movement, industrial worker strikes, and other movements created a explosive scenario before which the PRI government was too unprepared to produce adequate answers.

Second, Latin America as a geopolitical area had a remarkable visibility in domestic political terms. This visibility expressed itself in the milieu of Mexican progressive intellectuals and people politically committed around a vigorous support to the Cuban revolution, and a strong rejection against military regimes in Central and South America. On the one hand, it is well known that in Mexico the Cuban revolutionaries found support both openly and secretly, both in the beginning and years later, both morally and politically, and to a lesser extent, in financial aspects. Left-wing intellectuals and others who were politically engaged and active had taken the Cuban revolution as a motto and an emblem to their movement. On the other hand, U.S. occupation of some strategic sites (El Canal de Panama, Guantánamo), and the military regimes in Latin America (Nicaragua, Brasil, Paraguay, Uruguay, Chile, Argentina, El Salvador, Guatemala, inter alia) had produced a generalized indignation for the extreme forms of repression they had adopted, thus producing a clear enemy to fight against. Migrations of politically prosecuted people coming especially from Chile and Argentina, told the Mexicans about genocide, torture, missing people, inter alia, producing a narrative of horror thus easily producing the image of the antagonist, the crystal-clear representation of the type of regime that had to be eliminated at all costs.

Third, there was the condition of the nearness and visibility of Vietnam war and U.S. imperialism. Tracing back some historical lineage, it has to be remembered that when World War II was over it became evident that the world had changed in geopolitical terms. The United States of America managed to occupy the site of military, economic, political, and also to some extent, cultural power, however, not without contestation and resistance. The balance was then inclined not to European countries but to the U.S. predominance, with the Cold War at its center. From then onward a planetary political frontier was constructed between East and West, Socialism and Capitalism, national sovereign capacity or self-determination and non-asked for intervention, and so on. In addition, communication media emerged as a vigorous mean to interconnect and expose things happening all over the world. In this scenario, the Vietnam war was doubtlessly associated with Capitalism, interventionism, Western domination and U.S. imperialism. Due to the geographic contiguity between the United States and Mexico, this representation was shared by Mexican Left intellectuals and those politically receptive and involved. This also sets the conditions for the inclusion of Marxism as an ingredient of the progressive imaginary.

Fourth, the student movements in 1968 (Paris, Berkeley, Córdoba, Essex, Colombia, inter alia) had an expression in Mexico. Closely connected with the previous condition (i.e., Vietnam war), the student movement emerged showing a strong discrepancy against the establishment in its different incarnations (e.g., imperialism, colonialism, racism, double morals, bigotry, authoritarianism, inter alia). This movement spread throughout many countries articulating itself with specific scenarios in each site, and sharing, however, some representations of what had to be eradicated and excluded. In Mexico and—mainly but not only—in Mexico City, the movement was articulated initially around public university demands, and gradually started expanding its petitions to other democratic political questions, and thus achieved more supporters: not only higher education students but also university staff and even authorities, younger students, student's relatives, worker unions, professional associations, and even some peasant organizations. This growing movement paid the price of a pitiless killing in the massive demonstration that took place on October 2. Newspapers registered that the police and army indiscriminately attacked all those who participated in the demonstration backing up the movement (i.e., young students and their relatives, journalists and intellectuals, scholars, unionists, and party activists inter alia).

Fifth, there was the intellectual background that involved, on the one hand, tensions between Marxism and the theory of development, and on the other, the inscription of pragmatism as the philosophical antagonist to dialectical and historical materialism, and behaviorism as merely a Psychology approach.

Above I have held a non-reflective notion of context-dependent text.[30] Accordingly, I insist here that the conditioning form of the relationship between the four previous circumstances and the intellectual background I describe in this paragraph, is not that of a reflection of so-called external causes, but of an inscription, that is, a symbolic process by means of which the meaning of something in the context is engraved or sculpted on and penetrates a discourse thus becoming a part of its texture and altering the system of meaning as such.

One of the generalized forms of left-wing academic political involvement was closely tied with Marxism (both orthodox Marxism and its Latin American versions) since the almost nonexisting right-wing activism was hardly linked to any literature whatsoever.[31] An interesting feature of this setting is that both so-called hard and social scientists found in Marxism a source of inspiration for academic politics. In this context, theories of development by stages were linked with a sort of intellectual imperialism presenting the image of some "developed" countries as the model that were supposedly to be emulated. The humanities were also divided in a polarized manner. On the one hand, there was a radical and militant humanism, whose source of information and values was again Marxist literature, and on the other, Analytic Philosophy and Pragmatism. Finally, Behaviorism was invalidated in both the social sciences and in the few occasions discussed in philosophical debates as a "mere" psychological approach, strongly based on empiricist methods, structured around a highly simplistic epistemological view, and politically linked with devious ends (such as brainwashing).

In speaking about Marxism as a source of inspiration for progressive political, ethical, and epistemological values in those days, one has to include Marx's texts as well as many of the whole tradition that goes from the First, Second, Third, and Fourth International, to the less orthodox traditions; from the oversimplified handbooks and manuals edited both in the former USSR and in other nations (in Spanish e.g., Cuba, Chile), to deeper and highly complex philosophical, ethical, epistemological, and political discussions. In the 1970s, there was a proactive, dynamic, and sometimes feverish academic appropriation of Marxism (alternately discussing whether one should adhere to Lukacs, Althusser, or Gramsci, and experiencing how the "organic links with the proletariat" had to be constructed), as well as unionist interpretations in order to get acquainted with their so-called objective interests.

All the conditions presented above produce, among other things, a peculiar atmosphere in which the inscription of Dewey's ideas takes place. Differently from the first inscription commented earlier, this second articulation of Deweyean thought is not within a revolutionary discourse ruling the national educational system, but as the antagonistic link with an academic discourse arranged around Marxism, operating as a nodal point of political commitment.

Looking for citations of Dewey's work in the 1970s, several important libraries in Mexico City were consulted[32] and not many results were obtained. A search in the 1980s was needed in order to include ten more documents. From the thirty-one pieces registered,

- almost fifty percent were in Spanish and the rest in English;
- only seven dealt with education and the other twenty-four discussed philosophical matters;
- ten were written by someone other than Dewey, and twenty one by John Dewey himself.

In this search I organized the documents dealing with his work in three different types of editing formats: books and journal articles, teaching syllabi, and educational handbooks and dissertations.

Regarding books and articles, there were basically those written by Dewey himself, which means that his work was openly available as a primary source in libraries for students and scholars. Apart from this, books about his work were scarce and mainly dealt with philosophical issues.

Considering syllabi, Dewey is an ineluctable name in both Philosophy and Pedagogy; it is however, better studied and considered in the philosophical milieu than in the educational one.

Concerning dissertations and reference books, only a handful of pieces were found dated in the 1970s: mainly re-editions of handbooks of Philosophy, and the History of Pedagogy. In these it is a general rule to find his name contiguous to other remarkable pedagogues, however in a rather sketchy and shallow manner as is usually the case in this type of textbooks. Not a single dissertation about Dewey was found in the 1970s, and only three in the 1980s.

There was however, one M. Phil. dissertation in Educational Research dated in the late 1970s written by Adriana Puiggrós.[33] It is not about Dewey's philosophical or pedagogical legacy but concerns imperialism and education in Latin America, and mentions Dewey's perspective as a good example of the functionalist tradition in these countries. Because of its high academic standards, this dissertation was later published as a book and has been reprinted more than thirteen times.

This book has been acknowledged by many as a serious, well-informed, and profoundly analyzed piece of research. It constructs the Deweyean proposal as one of the sources of U.S. imperialist pedagogical discourse introduced to Latin American educational systems during the second half of the twentieth century as a means to ideological, political, and pedagogical unification.

Taking as references *Democracy and Education*,[34] *The Child and the Curriculum*,[35] and *Experience and Education*,[36] the author associates Dewey's notion of progress as the natural and intrinsic character of society, with the evolutionist concept permeating pedagogical functionalism.[37] She claims further in her book, that he took the inevitability of human evolution from Charles Darwin (i.e., from a naturalistic frame), and translated it into philosophical and educational language. In Dewey's conception the present is the development of past traditions, and stability amounts to the achievement of a better degree of equilibrium, which is reached by means of an evolutionary process. And when the time comes to make a choice between gradual progress and radical transformation, Dewey faithful to his Darwinean evolutionism, prefers gradual progress.[38]

In dealing with education and social class, Puiggrós links Dewey's curriculum schemes with the capitalist attempts to control the subordinate class by educational means and thus dissolve its class identity.[39] This topic is further elaborated in her book, when the author deals with the ideological role played by education and the way in which Dewey defends his prospect of schools as laboratories toward the gradual construction of a new society. Citing *Democracy and Education* she makes Dewey's logics visible as in how and why the individual judgments and emotions would become homogeneous with the emotional attitude of the group, thus legitimizing ideological unification. School as a means to control may not be entirely the same in the progressive tradition and in the reformist perspectives, however, similarities between the two can be found. She argues that Dewey, backing Roosevelt's New Deal accepts the combination of regulated monopoly, state prevalence, individualism, and institutional reformism. And it is in this field where control as a key function of schools, takes its peak and unfolds in different realms: massive ideological irradiation, the modification of behavior learned within social class and family environments, the dissolution of nefarious values that had been learned in social institutions and may be antagonistic with those promoted by school, and finally, the solution of political and pedagogical mediations in the didactic field.[40]

The Deweyean universal scientific model (understood by Puiggrós as equivalent with the functionalist goal of a universal civilization) is seen as a means to control the development of society toward capitalist progress without obstacles (i.e., the dream of theories of development in Latin America). It is represented by teacher and school as a means to dissolve extra school (vicious) experience. The educational laboratory would thus consist of rules and players. Quoting *Experience and Education*, Puiggrós remarks Dewey's centrality of experience as a means to lead the child toward knowledge already known by the expert (i.e., the teacher). This also means that the players (i.e., the pupils) may learn an order without noticing it is something externally imposed. This is possible as long as this order is collectively assumed, and does not come from an individual. School action is thus divided into a pair of forms: consensus and coercion that characterize the typical behavior in democratic life. Finally, the author claims, social engineering emerges as an outcome of the universal scientific model cherished by Dewey, however, in the imperialist imaginary, it achieves political overtones never imagined by its creator.[41]

Thus the assemblage/articulation of Dewey's ideas with imperialism, U.S. domination, and behaviorism is produced in the Mexican academic world of life (*Lebenswelt*). John Dewey is then constructed as a U.S. intellectual representative (*ergo*, he is associated with U.S. imperialism), and Pragmatism is understood as a sociological utilitarian and individualistic approach dismissive of ethical and political principles. A Marxist essentialist reading of Dewey takes place at this point. Puiggrós has indeed produced far more complex thinking criticizing essentialism in its different incarnations, however, this is another issue.

The inscription of Dewey's thought as an intellectual source for the revolutionary educational crusade (in the 1920s) or his undeniable links with the nonorthodox left wing in the late 1930s, seem to be erased from the horizon, and instead Dewey's project becomes an indisputable though somehow uncomfortable spot in the history of Mexican education.

The second Mexican educational reading of Dewey's writings enables the understanding of the ways in which his notions of democracy, progress, education, and the individual, no longer fit in the Mexican educational imaginary. Dewey thus becomes part of an antagonistic other, which through denial is excluded from the images shaping the modern self of the second half of the twentieth century.

Last Considerations

The turn of the century at a planetary scale, marked not the beginning but the rise of Liberalism as a feature of modern political thought in Western countries. It was equated with a progressive move. Pragmatism in philosophical terms, apart from Charles Sanders Peirce in the United States and (the second) Wittgenstein in Great Britain, was not really prevalent before Dewey's emergence. In countries like Mexico, Liberalism became predominant in the second half of the nineteenth century but it was represented as a political approach (rather than a philosophical perspective), and was articulated with Positivism, not with Pragmatism.

The comparison is made between two progressive discourses. In the 1920s it is a revolutionary program designed and implemented by the brand new power block, and in the 1970s it is a socialist academic discourse. Concerning the appropriation of Dewey in the 1920s it was amalgamated with the official educational program that was already inhabited by other pedagogical sources, and in the 1970s, the knowledge system does not have the same visibility compared to the political dimension of academic life (because of the nearness of 1968). So there is an asymmetry in the conditions in which Dewey is read and inscribed in these two progressive imaginaries.

I have tried to substantiate my claim that the Mexican inscriptions of Dewey's thought enable us to represent him as an *indigenous foreigner* based on two theses: His ideas of the individual inscribed in the fabrication of the modern democratic self in these two moments of Mexican history, and these two versions of progressive ideals are resignified in the very process of their articulation with Mexican discourses. In this process they are reoccupied and colonized thus fabricating a Dewey who is more and less than John Dewey.

On the one hand, the first moment commented earlier, indicates an inscription whereby Dewey's pragmatism in education, his approach to democracy, and its incarnation in the social laboratory represented by the School of Action, come to be a constitutive part of Mexican education in the postrevolutionary reconstructive period. Not to mention that, in addition, his perspective is inscribed in the very indigenous education schemes. The indigenous side of the inspiring expression "indigenous foreigner" does not depend on the literality of this fact, but rather on the complex symbolic process whereby the travelling (i.e., displacements) of Dewey's library (i.e., intellectual dispositive) stops momentarily[42] and gets fused with a salvation narrative (i.e., the Mexican revolutionary mystique). Dewey gets Mexicanized in becoming part of a chain of equivalencies[43] by means of which the horizon of plenitude and fullness is represented. His library comes to be a source without which the Mexican educational crusade is unthinkable, and thereby the assemblage of pragmatism and the other intellectual sources (i.e., Spanish Rationalism, Socialism, the New School, Liberalism, inter alia), with the peculiarity of Mexican scholars, educational officers, and schoolteachers understanding of Dewey's ideas, takes place.

This Mexicanization however, does not take place in the limbo. Rather, just like any other discursive articulation, it needs an outside, a constitutive other setting frontiers and demarcating what "it is." To put it briefly, this appropriation does not take place apart from an antagonistic other represented in this peculiar moment by Vasconcelos, the minister of education. Beyond historical actors that may occupy the position of antagonism, I am more interested in underlining the role played by Dewey's discourse in the fabrication of the modern self. I do not overestimate the position of Pragmatism in the educational crusade or in the promise of plenitude made by the Mexican revolution, however, education in the Mexican history of the twentieth century shows indelible marks, permanent indications of this inscription.

On the other hand, the second reading of Dewey's library presented earlier (i.e., during the 1970s) marks an entirely different form and quality of inscription. It is plausible to read this process as an emphasis on the foreigner side of the "indigenous foreigner," the metaphor that structures this whole book.[44] The foreign character can be read in the symbolic distance conveyed by the scarcity of research, essays, and publications produced in that period (in spite of the fact that there had previously occurred some positive adherence to Dewey's approach). This somehow means its partial exclusion from the research milieu and its positioning in a secondary status, basically in the teaching environment and not even as a subject in its own right, but just as one more name in the extensive list of Western pedagogues. The alien character can be observed also in the distance and disapproval with which his work is treated as an alien doctrine, representative of the very source of an academic perspective that is undervalued (i.e., functionalism), an ideological assignment that is objected to, and a political inclination that is doubtlessly rejected and denounced. In the 1970s, one comes upon another version of modern self, the one structured by one of the competing promises of plenitude: socialism. I am far from meaning that this second inscription is not overdetermined. My point is that this overdetermination involves other political, intellectual, and ideological ingredients. The expulsion of Dewey's Pragmatism from the cherished political and intellectual values in the progressive imaginary of the 1970s does not take place in an empty horizon. Rather, just like any other discursive articulation, it takes place in a political arena; however, contrary to the first inscription, it is Pragmatism that is occupying the position of the constitutive outside, and operates to negate what is supposed to be the narrative of salvation for the reference examined (i.e., Latin American socialism).

Considering the historicity of representations, both inscriptions lean on competing emblems of modernity (metaphysics and pragmatism, positivism and spiritualism, capitalism and socialism, Marxism and functionalism), and both involve their respective salvation narratives, and anticipated paradises. Accordingly both involve ingredients of the Enlightenment myth in which school education can bring about progress and happiness, and thus the fabrication of the modern self can be paradoxically traced in both inscriptions.

Dewey's library, as any other assemblage of intellectual sources, as any other political, ethical, and epistemic dispositive, is in turn merged into an already overdetermined field, crisscrossed by multiple forces and tendencies, different in kind and intensity. It eventually stops and gets fused with some of them to get later detached from the former and get fused into a different articulation. The political positions this library can occupy, however, indicate a primacy of the position over the literality of the text itself. There is no room for an essentialist reading of this traveling library.

Considering Dewey's circulation from a different geopolitical focus, the enterprise of gathering all these inscriptions in so different environments (i.e., Finland, Serbia, Japan, Sweden, Turkey, Brazil, Germany, the Soviet Union, Spain, China, Argentina, Switzerland, or Mexico) invite us to contemplate the family resemblance of this diversity, and obstinately iterate the questions: What Dewey? Why? When? How?

Notes

1. Accordingly I am not responsible for those peculiar assemblages (oversimplifications and politically led interpretations). Indeed I am responsible for my *own reading of these two inscriptions*.
2. Richard Rorty, *Achieving Our country* (Cambridge MA: Harvard University Press, 1998).

3. I do not understand discourse as merely linguistic signification, but as open-ended systems of meaning that involve both linguistic and nonlinguistic practices that structure social life (Ernesto Laclau and Chantal Mouffe, *Hegemony and Socialist Strategy: Towards a Radical Democracy* (London: Verso, 1985); something rather close to Wittgenstein's *language games* conceived of as the totality of words, objects, and action gathered around meaning, and representing forms of life, Ludwig Wittgenstein, *The Philosophical Investigationsm* (Oxford, Basil Blackwell, 1963).

4. Jacques Derrida, *Margins of Philosophy* (London: The Harvester Press, 1986).

5. This three-year school level comes after secondary and before higher education and is equivalent to the French Baccaleaureat.

6. Overdetermination stands as an objection to deterministic and teleological causality (fixated), and instead it involves a double discursive operation: displacement (i.e., circulation or traveling) of meaning throughout different signifiers, and condensation (or the temporary fusion and articulation) of these meanings with others at a certain point. It also entails that no identity is pure or uncontaminated, but rather has in it the traces of other identities (i.e., it is relational and not positive. See Rosa N. Buenfil Burgos, "The Mexican revolutionary mystique," in *Discourse Theory and Political Analysis. Identities, Hegemonies and Social Change*, ed. D. Howarth, A. J. Norval, and Y. Stavrakakis (New York: Manchester University Press, 2000).

7. Allan Knight, *The Mexican Revolution* (New York: Cambridge University Press, 1986).

8. Buenfil Burgos, "The Mexican revolutionary mystique."

9. *Malinchismo* is a rather unfortunate Mexican word meaning the overvaluation of the foreign and underestimation of the native, which became a constitutive feature of Mexicans during the colonial period, and albeit it's being bitterly criticized it has not been eradicated.

10. Modernism as an abstract value has been present and representing a desire and a horizon of plenitude already, at least, since the second half of the nineteenth century in Mexico. Thus it is not that the revolution brought it about, but it is reasonable to say modernism to be synonymous with a horizon of plenitude (if something were modern it would be good to have it).

11. Gamio and Sáenz met Dewey at the University of Columbia where both studied their Ph.D: Gamio in Anthropology and Saenz in Science and Philosophy (see Rafael Ramírez, *Escuela rural mexicana* [México City: SEP80-Fondo de Cultura Económica, 1981]).

12. See Roderic Camp, *Intellectuals and the state in Twentieth-century in Mexico* (Austin: The University of Texas, 1986); Mary Kay Vaughan, *The state, Education and Social Class in Mexico 1880–1928* (Illinois: Northern Illinois University Press, 1982); Ramón E. Ruiz, *Mexico. The Challenge of Poverty and Illiteracy* (San Marino, CA: The Huntington Library, 1963); Rosa M. Torres Hernández, "Influencia de la teoría pedagógica de John Dewey en el periodo presidencial de Plutarco Elías Calles y el Maximato 1924–1934" (Ph.D Thesis, Mexico, Universidad Nacional Autónoma de México, 1998).

13. See e.g., the splendid collection of pictures produced by the Fondo Casasola and Ortiz Monasterio Pablo, ed., *Jefes, héroes y caudillos* (Mexico: Fondo de Cultura Económica, 1986).

14. John Rutherford, *An Annotated Bibliography of the Novels of the Mexican Revolution of 1910–1917* (in English and Spanish, Troy, NY: Whitston Publishing Co., 1972).

15. Carlos Pellicer and Rafael Carrillo Azpeitia, *Mural Painting of the Mexican Revolution* (México: Fondo Editorial de la Plástica Mexicana, 1985).

16. See Hernandez, "Influencia de la teoría pedagógica de John Dewey," 23.

17. The New School already was a hybrid formation composed by different pedagogical theories and educational programs such as Abbotshome and Bedales, Montessory, Decroly, Kershensteiner, Parkhurst, Wirt, and Washburne, inter alia (See Hernandez, "Influencia de la teoría pedagógica de John Dewey," 195, footnote 25).

18. José Vasconcelos, *Antología de textos sobre educación*, ed. A. Molina (México City: Fondo de Cultura Económica, 1981).

19. José Vasconcelos, *Antología de textos*, 48 my translation, see especially pages 54 and 67.

20. Both documents by Ramírez 1924, "La Escuela de la Acción en la educación rural" and "La Escuela técnica para campesinos," in Rafael Ramírez, *Escuela rural mexicana* (México City: SEP80-Fondo de Cultura Económica, 1981).

21. Rafael Ramírez, 1935 "*La escuela proletaria: Cuatro conferencias sobre educación socialista*," in Rafael Ramírez, *Escuela rural mexicana*.

22. Rafael Ramírez, 1937, *Técnica de la enseñanza* (México City: Escuela Nacional de Maestros, 1970).

23. In the thirty written pieces registered by his biographers from 1928 to 1949, the titles of Ramírez's texts suggest an articulation of Dewey and Kropotkine in dealing mainly with rural school, proletarian education, and didactics for specific subjects (i.e., history, reading, mathematics, inter alia).

24. John Dewey, "Interest in Relation to Training of the Will," in *The Early Works of John Dewey 1882–1889*, Vol. V, 1895–1898, Introduction by William McKenzie (Illinois: Southern Illinois University Press, 1975).

25. John Dewey, "A Pedagogical Experiment," in *The Early Works of John Dewey*.

26. Moisés Sáenz (1888–1941) was more focused on policy and the administration of indigenous schooling, and his writing from 1926 to 1964 is mainly about indigenous education as a means to incorporating them into the national milieu, and integrating civilization into them.

27. John Dewey, *The School and Society: Being Three Lectures by John Dewey Supplemented by a Statement of the University Elementary School* (Chicago: University of Chicago Press, 1907).

28. Moisés Sáenz, "Mexican integration through education," Conference imparted in the University of Chicago, 1926 in Moisés Sáenz *Antología de Moisés Sánez*, ed. Aguirre Beltrán (México City: Ediciones Oasis, 1970), 17.

29. See Rorty, *Achieving Our country*.

30. Also in Rosa N. Buenfil Burgos, "Constructions of the child in the Mexican legislative discourse," in *Governing the Child in the New Milennium*, ed. by

Kenneth Hultqvist and Gunilla Dahlberg (New York and London: RoutledgeFalmer, 2001).

31. These few groupings consisted of front men and body guards, groups of male cheer leaders and in the best of cases, well-intended religious youth.

32. Ten databases were visited (three at the National University) and the rest in other important universities or research centers (Instituto Mora, El Colegio de Mexico, Departamento de Investigaciones Educativas, inter alia) and twenty-one texts were concentrated in one library, and none of these texts was written by a Mexican author. In two other databases ten documents were found but had been edited in the 1980s.

33. Adriana Puiggrós, *Imperialismo y educación en América Latina* (México City: Nueva Imagen, 1980).

34. John Dewey, *Democracy and Education. An Introduction to the Philosophy of Education* (New York: The Free Press, 1916).

35. John Dewey, "The child and the curriculum," in *The School and the Child; Being Selections from the Educational Essays of John Dewey*, ed. J. J. Findlay Publisher (London: Blackie & Son Limited, 1906).

36. John Dewey, 1937, *Experience and Education* (New York: Collier Books, 1963).

37. Puiggrós, *Imperialismo*, 16.

38. Puiggrós, *Imperialismo*, 104, 105.

39. Puiggrós, *Imperialismo*, 33.

40. Puiggrós, *Imperialismo*, 107, 150.

41. Puiggrós, *Imperialismo*, 228.

42. Indigenous evokes the idea of something having originated in and being produced, growing, living, or occurring naturally in a particular region or environment (innate, inborn, native), *Webster's Ninth New Collegiate Dictionary* (Springfield, MA: Merriam-Webster INC, 1990), 614

43. Laclau and Mouffe, *Hegemony and Socialist Strategy*.

44. Foreigner evokes the concept of a person, idea, or object belonging or owing allegiance to a foreign country, situated outside one's own country; alien, not connected or pertinent; occurring in an abnormal situation in the living body and often introduced from outside (see *Webster's Ninth*, 483).

9

John Dewey through the Brazilan Anísio Teixeira or Reenchantment of the World[1]

Mirian Jorge Warde

Overview

Between the 1920s and the 1930s, Anísio Teixeira is committed to inscribing the name of John Dewey as a necessary reference to the renovators of education in Brazil and he was determined to put Dewey's works in circulation.

Anísio Teixeira recorded, in several writings, that his encounter with the North American culture and education, and especially with John Dewey's pedagogical and social ideas, represented a definite break with his previous life and thinking about social and educational problems. Teixeira's writings from the 1920s and 1930s indicate his effort to make it evident that the encounter with the United States positively filled the gap left by Europe, which disappointed him regarding social alternatives and, even worse, regarding the orientations of philosophical thinking, namely that of Catholicism.

The analyses developed in this chapter focus on: (a) the mental operations that Anísio Teixeira made to appropriate Dewey's ideas and put him into circulation; (b) the political and cultural ambiences in which Teixeira circulated and which put Dewey into circulation, and (c) the local intellectual networks, both domestic and international, which Anísio Teixeira was a part of, where his social belonging allows him to identify himself with the mental configuration that built the Dewey-form, both for his own public and private use.

Introduction

In Brazil, Anísio Teixeira (1900–1971) is considered to be the major Brazilian representative of Dewey's thinking and his most systematic disseminator.[2] Since the 1920s and the 1930s, Anísio Teixeira is committed to inscribing the name of John Dewey as a necessary reference to the renovators of education in Brazil, especially those who participate in the so-called new school movement, and he was determined to put Dewey's works in circulation, by translating them into Portuguese, which he mostly did himself.

Tieing and untieing John Dewey in different social networks, Anísio Teixeira inscribed him in the Brazilian educational milieu, like it happened in other countries, as an indigenous foreigner.

Anísio Teixeira recorded, in several writings, that his encounter, in those decades, with the North American culture and education, and especially with John Dewey's pedagogical and social ideas, represented a definite break with his previous life and thinking about social and educational problems. This encounter represented, indeed, the recovery from a profound spiritual crisis.

Teixeira's writings from that time indicate his effort to make it evident that the encounter with the United States positively filled the gap left by Europe, which disappointed him regarding social alternatives and, even worse, regarding the orientations of philosophical thinking, namely that of Catholicism.

During the 1940s and 1950s, Anísio Teixeira sustained the same theses regarding the intimate association between the cultural and educational North American standards and John Dewey's social and educational thinking. In the 1960s, however, a number of reasons made Teixeira pessimistic about the new trends of North American society and culture while his belief in Dewey become more solid.

Thus, the period when Anísio Teixeira was first in contact with the United States and with Dewey's ideas and the final period just prior to his death, in 1971, represent two moments of John Dewey's presence in Brazil. It is the presence of Dewey not as a person's name—an author's name—but as a representation of the North American lifestyle and a modern way of thinking social and educational issues. One could say that little read although very quoted and discussed, John Dewey has been converted into a prototype of a certain society and education.

Such a John Dewey formatted by other hands—Anísio Teixeira's hands—is here called the Dewey-form.[3]

The analyses developed in this chapter focus on: (a) the mental operations that Anísio Teixeira made to appropriate Dewey's ideas and put him into circulation; (b) the political and cultural ambiences in which Teixeira circulated and which put Dewey into circulation, and (c) the local intellectual networks, both domestic and international, which Anísio Teixeira was a part of, where his social belonging allows him to identify himself with the mental configuration that built the Dewey-form, both for his own public and private use.[4]

A Man and His Dilemmas

What was happening in North America in terms of society and education began to capture Anísio Teixeira's attention in the mid-1920s; attention to the social and educational ideas of John Dewey came as a result of a broader interest. Teixeira formulated at once the notion that Dewey's thinking was the master key to understanding the North American social and educational trajectory; J. Dewey would be the purest substratum of that society.

Europe and Catholicism were the first references in Teixeira's mental configuration. As counterfaces to such references, he interpreted the signs of North American society plus he read and re-wrote Dewey.

Teixeira came from a profoundly Catholic family for whom the calling to the clergy was thought of exactly as the calling to political life under the shadow of his father, an important political leader in the local community (the state of Bahia, in Northeast Brazil). These two destinations outlined the field of Teixeira's first intellectual and professional choices.[5]

The high Jesuit clergy, who directed two schools of the Society of Jesus in which Teixeira had taken his primary and secondary studies, invested in his education in a special way; they expected to draw him to the religious life and to dedicate himself to elevated intellectual missions.[6]

Anísio Teixeira's religious calling was relatively well manifested by the end of secondary school, and his parents had been consulted about letting him continue his religious education so that he would become a member of the Jesuit Order. As his parents vetoed the idea, Teixeira followed his father's advice to study Law, although he had no special interest in legal activities.[7]

The correspondence between Anísio Teixeira and his family members and with his main spiritual mentor, Father Cabral, dated 1921, when he was about to finish Law school, indicates that appeals to the religious life had not ceased.[8] This was the main subject, while the issues involving his legal education did not concern him.[9]

However, doubts assailed Anísio Teixeira, undecided between a mundane career and a spiritual career, the private retirement and the public engagement. Those doubts unfolded into other issues concerning the possibility of finding spiritual runways that would not disconnect him from the world.

In the early 1920s, when writing about himself, Teixeira speaks of his worries involving the Catholic religion that provided him with solutions for life, but which in fact meant a very renunciation to life.[10] In those years, Teixeira seemed embarrassed to mobilize traditional Catholic solutions, unaware of the new modernizing formulas worked out by different sectors of the Catholic intelligentsia and by the movement known as the aggiornamento unleashed by the Vatican.[11]

Curiously, his spiritual mentor, Father Cabral, had already told Teixeira to view the world from a distance so that he would not suffer. In a letter dated April 1921, he preaches to his former student:

> Courage, my good Anísio! Consider yourself as of now as a religious person who is forced to live among mundane people; sanctify yourself more and more, do around you all the good you can; enhance with more and more frequent acts the wish to leave the world and trust that the difficulties of your parents, the Heavenly Father can solve at any moment.[12]

But it was not only the traditional Catholicism of the Portuguese Jesuit fathers that embarrassed Anísio Teixeira; with equivalent power, the traditional and clientele-like way of doing politics also circumscribed his field of action.[13] In 1924, complying with his father's pressure, he went to see the newly elected governor of the state, to ask for "a modest position as public attorney" in his hometown. Instead of responding affirmatively to the request, the governor offered him the highest position of director general of public instruction in the state of Bahia as a way to reward the Teixeira family for their support in the election.[14]

At that time, according to Teixeira's own account, the governor gave him a book by the Belgium writer Omer Buyse, *Méthodes Américaines d'Éducation Générale et Technique* (1908, 1909, 1913),[15] so that he could get started on the educational issues about which he had no previous training and he had never before given any attention. Buyse's book would provide Teixeira with the first opportunity to get in contact with Dewey's proposals in the educational field.[16]

He recorded, at the time, the impression Buyse's book had provoked in him; it had revealed to him a new way of looking at education and stimulated him to enter a period of intellectual relentlessness and revision because it had undermined his old beliefs.

The book had impressed him exactly because it had shown him a victorious society, from the economic and social point of view, which threatened European supremacy and was the fruit of the North American entrepreneurial spirit, the results of family and schooled education. This education was based on the requirement of an active and personal initiative and effort. In short, education was the key to North American prosperity.

For Teixeira himself, it was amazing that this analysis had been made, early in the century, by a Belgium educator who denounced the process of pauperization and ineffectiveness of the European school education as a result of its prejudices and the ignorance regarding the advancement of the science of education. What Buyse had revealed to him was an educational system that functioned well and in perfect integration with modern society; for these reasons it was a successful system.[17]

In 1925, despite being very impressed by what he had become aware of in terms of North American education, Teixeira decided to take his first study and observation trip to schools in Europe. He chose France and Belgium, and the city of Rome, to take part in a great international Catholic event, promoted by the Vatican.[18]

Teixeira wrote in his diary that he was left full of doubts and came back to Brazil plagued by a huge "spiritual crisis." The modernity he had seen in Paris left him profoundly upset. All in that city seemed fake, artificially added; the artificial lights of the city hid the spiritual darkness into which modern man had submerged. Not even the visit to the Vatican, the papal blessing, calmed his soul: the effort that the Catholic Church was making to update itself seemed to him insufficient to face what was not only a personal crisis but mainly the spirituality crisis of modern man.[19]

He also recorded his subtly positive impressions about the monarchical regimen, with open praises to the bucolic ambience in rural areas of Portugal—which he had visited on his way back to Brazil—, in contrast to the moral harms resulting from the institution of the Republic in that nation. Particularly curious are his comments about the optimistic horizons projected by the new French intellectuals who demanded leadership from the Catholic Church in an aimless world and a mankind that was morally and intellectually disconnected; the intellectual ambience in Paris contrasted, therefore, with the advanced material and technical ambience of the city.[20]

On the other hand, in an interview to a newspaper, Teixeira says that, in his visit to France, he had worked to strengthen old admirations for the French civilization and culture. Paris, which had frightened him with its material seductions, required however his deepest attention because in that city the highest Catholic intellectual circles could be found. There lived Jacques Maritain and Charles Maurras, around whom Paris impelled entirely "an intellectual youth profoundly willing to free itself from the powerful strands of skepticism, philosophical or political rationalism." Those thinkers replaced "the essential truths of authority, order, balance and hierarchy which are indissolubly connected with the religious truths and with the Catholic truths and welcome the Church as the most beneficent of the social forces."[21]

Anísio Teixeira was in Paris exactly in the decade when the city was going through its major intellectual effervescence between the two world wars. The short period in which he stayed there had been enough to sharpen his senses to the new waves that were shaking the old philosophical structures. At that time, he surrendered to the criticism he had received, more than a year before, from a Catholic intellectual with who he had circulated, that he suffered from "a century of mental decay" because he was still fond of Thomist ideas when the major philosophical event of the moment was taking place in Paris; according to the author of the criticism, Maritain was responsible for the "revival of the philosophy" of Saint Thomas Aquinas.[22]

There Are Many Ways that Lead to Salvation

As soon as he took over the position of director general of public instruction, in 1924, Anísio Teixeira read Omer Buyse and contacted educators who were, in several Brazilian states, reforming their educational systems by adopting measures intended to modernize school organization and pedagogical practices. In his early days as a manager, Teixeira got to know some leaders who a few years later would draw him into the epicenter of the so-called new school movement.[23]

Dealing with educational matters on a preliminary basis, Teixeira became engaged in versions of what North American modernity would be, all of these worked on by Europeans or Brazilians. In the first of these versions, Teixeira began to construct his own version of what he saw as modernity, linking it to the North American standards against the "obsolete" European or the "primitive" Brazilian standards.[24]

The first contacts with the new educational reform trends did not pressure Teixeira to discard his religious beliefs, both because most leaders of the educational renewal were Catholic militants and because Teixeira had not been provoked into thinking about new educational standards in relation to religious choices.[25]

While he was being introduced to the educational "modernity" in North American terms, Teixeira was also making contact with the new tendencies of European Catholic thinking. In this case, however, Teixeira realized that it was not modernity; on the contrary, the way the Catholics reacted to the new social ambience—Bourgeois, industrial, urban, and so on—was assimilated as reactions against modernity, that is, they were the antimodernity or non-modernity.[26]

Dissociating "ideas" from "reality," Teixeira worked both to build his understanding of the new philosophical trends in the Catholic milieu and to produce his own version of the North American education standards and also, to construct, a little later, the Dewey-form.

Anísio Teixeira did not question the prevailing Catholic version which claimed that the process of aggiornamento inaugurated by Pope Lion XIII after the publication of the Encyclical Rerum Novarum in 1891, as well as the new philosophical tendencies of the Catholic intellectuality, were instituting modernity in spite of declarations to the contrary.[27]

Resisting the bourgeois order and reacting against Capitalistic materialism was the discursive strategy effectively adopted by the Catholic Church and its intellectuals to insert Catholicism into the new state of affairs and, at the same time, declare its perpetuity, that is, its transcendence in relation to the mundane life. By disregarding human time and space in favor of supernatural uchrony and utopy, the Catholic Church official doctrine reaffirmed its holiness and its disdain of material things as well as its scorn for political disputes.

Since the mid-nineteenth century, there is within the Church what one may call a theological crisis arising from the need to resolve the relations between science and faith.[28]

With regard to the social field, from the hierarchy of the Church and the intellectuals who spoke on behalf of the Catholic faith, a wide spectrum of positions has been formed, ranging from the Millennium believers—who refuted Capitalism and the Bourgeois order in favor of an idyllic return to the life of the Middle Ages—to the Catholic socialists—who claimed Catholicism to be the origin of a nondegenerated socialism, that is, the socialism that subordinated material reform to moral reform.

The Vatican attempted to anchor the official positions of the Church to an equidistant point between the regressive excesses and the reforming excesses. It declared that the Western world had plunged into a profound spiritual crisis, which made it urgent to implement reforms based on moral conversion—to be conducted from individuals toward society—with the aim of constraining the materialisms imposed by capitalism, such as the greed for profit and the fondness for material values in detriment of spiritual values.[29]

None of the fighting fronts, occupied by the Catholic Church in different moments, was abandoned. In its aggiornamento, the Church managed to identify its enemies and draw among them lines of continuity: after the Reformation there came the Bourgeoisie and capitalism out of which both positivism and historical materialism and pragmatism sprang, upon whose shoulders rested both the Soviet and North American societies.

The papal Encyclical dated 1891—*Rerum Novarum. On the Condition of the Working Classes*[30]—represented the first and matrix-type Catholic response to the modern world. From front to back, it refutes the 1848 Manifest of the Communist Party, written by Marx and Engels. With the Encyclical, the Vatican summit vigorously opposed the Communist project and patented Catholicism as the superior modern project for modernity.[31]

The work of the two French intellectuals mentioned by Anísio Teixeira in 1925 are remarkable expressions of the different Catholic strands that were on the scene since the late nineteenth century, which the Vatican summit, on one hand, consecrated—in the case of Jacques Maritain—and, on the other, condemned—as was the case with Charles Maurras. Apparently, Teixeira did not manage, at the time, to distinguish the significant differences between them.[32]

One cannot find in the writings of Teixeira any record that he had attempted to examine the secular vector contained in the spiritualized discourse of the Catholic intellectuality. In contrast, a little later, Teixeira seems not to have been tempted to decipher the religious appeal he had found in Dewey's pragmatism.

The Encounter with the Prophet and His Kingdom

In 1927, he decided to take a new observation and study trip to the United States. It seems that at that time he had made up his mind to test another way to find the solution to his dilemmas.

The first impact from reading Omer Buyse's book had eventually produced its more direct effects. The visits to the European schools had not left impressions in the policies he was implementing in the public instruction of Bahia; apparently, it confirmed to him that the "new" in matters of education was no longer on that side of the world, because it was concentrated in several renewing initiatives in the United States.

At last, Teixeira decided to live his own experience of the new world. Instead of a Catholic priest or a Belgian educator, Teixeira would rather travel on his own, preparing himself "to visit the American country" by reading Henry Ford's book, *My Life and Work*.[33]

The feelings the book caused in him were intense: "In several fields—in spiritual life, one; in industrial life, another;—two books provided me with the feeling of fullness, of profound agreement, of the inexistence of doubts—The spiritual exercises by Ignácio de Loyola and *My Life and Work* de Henry Ford.". . . In the two works he found "this essential and unbreakable realism, which defines truth and gives it this force of coup, of thrust, which such books communicate to us."[34]

Comparing Ford's book with Loyola's is symptomatic; although he declared himself disappointed with his old mental mold and, as a result, he was existentially committed to finding a new way of facing social and individual problems, Teixeira molded his representations of North American society, education and pragmatist philosophy by opposing them to his representations of Catholic European society, education, and culture. Hence, the mental operations he made to take over the "new" always implied in submitting it to the old, to what was already known.

Ford's touchstone was "industry, understood as service, as a servant of the collective welfare (. . .) How far we are from the philosophical or socialist ideological thinking."[35]

Before making comments on Ford's book, Teixeira had already announced that his trip to the United States was driven by a certain sense of spiritual, religious search. As if he had repeated his pilgrimage to the Vatican, in 1925, just after reading the Bible. But in Ford he seems to have found more directly, with less suffering, the sense of usefulness that can be applied to things "with more intelligence, more wit."

It would seem that Ford provided him with a more honest and direct understanding of human weaknesses, channeling them into a positive meaning: he does not have a foolish faith in human perfection; there is confidence in saying that the worker will have a higher salary, the capitalist will have more profit, the public will have better service—and all will be happier.[36]

The way Anísio Teixeira understood Ford's book points to a dual direction: on the one hand, he counterposes the Catholic religion, which requires a great deal of sacrifices so that the prize can only be enjoyed after death, to this mundane, material, tangible religiosity of which Ford was the most complete expression.

> There is in Ford a lot of this evangelical confidence that there is room for all in the world. Each one must fulfill their duty and all will be given one hundred for one.
>
> [. . .]
>
> With no mention of God, the entire book breathes His Spirit, it is impregnated with order, abnegation, humility, subordination of man to anything greater than him.
>
> Far away from the divinization of man, in no work is he more subordinated, more connected, more in his own place. And these ideas are proven by the most overwhelming material triumph that one has ever seen.
>
> For me, there was no need for a greater proof of the magnitude of Ford's work: it is simple. There is no complication, just like the Gospel.[37]

On the other hand, Teixeira finds in Henry Ford the antidote for the revolution, for the social overturns that he learned to reject either in his political ambience or among the Jesuit priests. The touchstone, once more, is industry. Through it, Teixeira believes, men would be at the eve of suppressing misery [. . .], of a solid collective welfare, if men open their eyes to the examples of Ford's work; [. . .] because it will be through industry that we shall reach our material salvation, industry is only now in a position to bear up under the Ford dictum.[38]

Teixeira foresees opposition to Ford's methods, due to ignorance and stupidity, but these also would be redeemed by the lucidity of modernity. And even better: "with no revolutions. With no apprehensions. No overturns. Within the current regime and current order of things. With continuity, with advancement, with development . . ."

There it was, for him, "the secret of Ford's truth: we do not advance by leaps. And there is no leap in Ford's work. The no change of atmosphere. His work breathes our air. There is no destruction: there is taking advantage of; there is order; there is Clairvoyance."[39]

Teixeira ends his comments in a state of grace:

> That's why I did not finish reading the admirable book—without feeling that it obliged me to—the broader, the most confident and

the most generous act of faith in human will and work that has ever been done.[40]

The encounter with Fordism was, for Teixeira, a way of life in which religion could be secularized by investing energy in the production of "common goods." The encounter with Ford's concept of industry—which is much more than the factory because it implies the organization and control of the entire social and individual life by rationalized means—cemented his adherence to "Americanism" as a revolution of customs, a new civilizing cycle, that was shaping new forms of thinking, feeling, and living, to become the parameter of progress, happiness, welfare, and democracy. It was planting in the hearts and minds the silhouette of the "new man"—a rational, administered, and industrious being.[41]

After four months as an intern in the United States, Anísio Teixeira confessed to have pacified his spiritual crisis. Upon returning, he had converted to a new faith; the faith that no longer dwelled in a particular theology; it now had a steady base: the struggle for the educational cause. Thus education would have saved him from becoming disenchanted with the world, something he feared so much;[42] Teixeira clung to the educational cause giving it a universal salvational meaning.

On the trip he had taken, he was particularly affected by the information about the importance of John Dewey's philosophy on the principles and modern methods of teaching adopted by the schools, and, moreover, by the construction of a renewing pedagogy. When he came back to Brazil, he recorded that "John Dewey is, in America, the philosopher who most acutely traced the fundamental theories of American education. No other thinker has been given such an outstanding place in the systematization of modern educational theory."[43]

Even before making Dewey's work available in Portuguese, Anísio Teixeira included his ideas in the debates around educational reforms in 1928. Through his report on his United States trip which, although it had been sent directly to the state governor, seems to have been formatted for wide dissemination. It was the first gesture to install John Dewey as a traveling library in Brazil.

Under the title *Aspectos Americanos de Educação* (American Aspects of Education), he dedicates part of the account to what he observed in the schools he had visited and another part entirely dedicated to the presentation of the "education essentials," in which he summarizes "as faithfully as possible the ideas with which John Dewey presents the current meaning of education."[44]

His "conversion to a new creed," as he himself referred to his adhering to Dewey's philosophy was so decisive that Teixeira decided to retire from the General Board of Public Instruction in Bahia so that he could stay at Teachers College in Columbia studying for a year close to John Dewey. His return had already been arranged with the board of the International Institute in the College.

Anísio Teixeira was in Teachers College in the school year of 1928–1929, where he attended classes by William Kilpatrick and George Counts.[45] Unfortunately, he did not manage to attend any course by John Dewey or to meet him personally; Dewey was away from Columbia University at that time because of a long international trip in which he went back to Turkey, to China, he visited the USSR and Mexico. Teixeira certainly regretted not meeting Dewey, but the reasons were out of his reach; the same did not happen with Edward L. Thorndike and his team, from whose courses Teixeira intentionally kept his distance.[46]

Teixeira recorded with his class notes that, beyond all blemishes, that in the United States education towered over the "craziness for the dollar, for the material magnitude" and which did not forget about the "spiritual values of life." In another page, he wrote down: "America is reinterpreting life" by turning "education into a religion". . . by means of "the scientific spirit, pragmatism . . ."[47]

Between 1928 and 1929, the United States was under the perverse effects of the economic depression, but Teixeira left no records of his impressions on the social atmosphere he had found there. It is likely that Dewey's creed (to which he had adhered as the faith of a newly converted Christian) acted as an attenuating filter for the severity of the social problems he observed.

Teixeira returned to Brazil in 1929, with the title of Master of Arts; he was fully convinced about Dewey's pragmatism. The quick conversion to the new creed apparently was not due to the frugality of his former Jesuit education; Teixeira needed the holy amidst the world of men. He called for more spirituality in the modern world.

Upon his return he was certain that in Dewey's philosophy there was a successor of that transcendentality he had claimed in prior years because he no longer found it in Catholicism and in the European philosophies. Out of the new Catholic philosophical strands it seemed certain that he could not draw upon a modern sense of social engagement.

Teixeira understood that pragmatism was a peculiar philosophy when he infused a transcendental meaning to experience and at the same time he conceived reason as part of the immanence, that is, as an intrinsic part to the being of experience. Dewey's pragmatism, which he would rather call "instrumentalism," was a call to life lived as an experience, about which intelligence had to work on to create the future: that was the only metaphysics that could coexist with modernity.[48]

In his writings, Teixeira did not leave any record showing he had dedicated himself to understanding his pacification in Dewey nor why in that thinker he discovered the prophet who pointed him toward the new Jerusalem.[49] Had he conducted archeological studies on the land where John Dewey built a new spiritual dwelling for human beings, perhaps he would have understood the intricate connections between his old European and Catholic references and his new, North American and pragmatist ones.[50]

It was, however, through W. Kilpatrick and George Counts that Teixeira acquired new tools to shape the Dewey-form both for his own personal use and for public dissemination; from his teachers at Teachers College, he developed the most simplified and easy-to-grasp version of Dewey's pragmatism, depurated from its complex incursions in theology, philosophy, psychology, and child study.[51]

Teixeira studied Dewey's writings in which educational themes are central and in which Dewey does not give account of the philosophical issues that involve the members of the Metaphysical Club, the philosophers from the Universities of Harvard, Johns Hopkins, and Chicago, by considering a philosophy that would be able to take into account lived life.[52]

Having avoided the conceptual paths taken by Dewey, Teixeira cleared him from his intellectual network consisting of G. S. Morris, C. S. Peirce, G. S. Hall, as well as W. T. Harris, Josiah Royce, W. James, G. H. Mead, and others.[53] Not even the links beteween Dewey and the American movement in progressive education, were clarified; the Dewey-form put into circulation in the Brazilian educational milieu was above all that of a mentor for the "new school" as Teixeira claimed for Brazil.

The procedures to untie John Dewey from his native social networks composed the same body of operations to bond his name to the intellectual and political networks that Teixeira was joining.[54]

The first initiatives to put Dewey's ideas into circulation in Brazil (between 1928 and 1934) give account of the inaugural acts to constitute the form that Teixeira would boost in his successive rewritings. At once, Dewey was converted at the same time into an heir to the high

Western culture and an exclusive thinker; he was granted the prime honor of having assimilated Froebel, Hegel, Rousseau, and Herbart just to overcome them afterward. From Stanley Hall he would have learned to focus attention on the child, but it was on his own that he developed a philosophy and a pedagogy of how and for what reasons to educate such a child.[55]

To Dewey's partners at the Universidade of Chicago (1894–1904), Teixeira did not devote any special attention; neither for credit nor for discredit, he wanted to know from George H. Mead, James H. Tufts, James R. Angell, Edward Scribner Ames, Addison W. Moore, Jane Addams who were directly involved in converting the Philosphy Department whose head was Dewey into the epicenter of pragmatist studies.[56]

On the other hand, Teixeira outlined around Dewey a sanitary belt to protect him from real or alleged partnerships: of the relationship with William James he erased almost all traces, given his pragmatist excess; of the presence of C. Peirce he left no clues; E. L. Thorndike was discredited, without even being read, as the author intended to subordinate the soul to tests and measurements.[57] Teixeira also erased in Dewey his first philosophical education in which his masters had included Spinoza, Kant, and Hegel.[58]

Teixeira never denied the philosophical heritage that had been deposited in Dewey, but he did not take it into consideration while rewriting his mentor. His traces show in several texts, the scribbles that survived under Teixeira's writings.

The Dewey-form was thus the result of many depurations and few affiliations or alliances. More than what he did to himself, Teixeira perpetuated in Dewey the feature of the intellectual who had dispensed with almost all individual or group intermediations in order to compose himself.[59] Dispensing with all intermediaries, Dewey had made direct contact with the deepest sap of Northn American culture, and this allowed him to be, according to Teixeira, simultaneously his mirror and his most distinguished critic; his affirmation and his denial—his most finished synthesis.

In the written version, the Dewey-form is inscribed in Teixeira's personal texts—his diaries, correspondences, and interviews; in articles and books he signed; in educational policies he developed in a number of public offices in which he served as well as in the translation of Dewey's writings, which he did or gave his official approval.

Usually accompanied by prefaces or lessons to provide the reader with safe guidance, there are no more than nine books by John Dewey published in Portuguese by Brazilian publishing houses, out of which Teixeira translated five and took care of the publication of the other six (between 1930 and 1974): *Vida e educação* (Life and Education [1930]), *Como pensamos* (How We Think [1933]), *Democracia e educação* (Democracy and Education [1936]), *Liberdade e cultura* (Freedom and Culture [1953]), *A filosofia em reconstrução* (Philosophy under reconstruction [1958]), *Natureza humana e a conduta* (Human Nature and Behavior [1956]), *Teoria da vida moral* (Theory of Moral Life [1960]), *Experiência e educação* (Experience and Education [1971]), and *Experiência e natureza* (Experience and Nature [1974]).[60]

Only *Life and Education*—whose first edition is 1930 and the last reprint there is any record of is 1979—had considerable dissemination in the Brazilian educational milieu, due to the presence in the bibliography of teacher and pedagogy education courses, at least until the early 1970s.

The first three books were published in editorial series under the control of two leaders of the "new school," Lourenço Filho and Fernando de Azevedo; in the 1930s, Anísio Teixeira played the role of editorial counselor for both these men. Together, they decided to include Dewey's titles among the authors they had criteriously selected for the purpose of forming a renewed pedagogical culture. Among the first titles of the series coordinated by Lourenço Filho, in which *Life and Education* is a part of, one will find *Education of a Changing Civilization* by W. Kilpatrick, *Introdução ao estudo da Escola Nova* (Introduction to the Study of the New School) by Lourenço Filho himself and *Education and Sociology* by E. Durkheim. In the other series, Dewey appeared among the books by Fernando de Azevedo, Anísio Teixeira, and E. Claperède.

All in all, none exceeded the number of copies and reprints achieved by the tiny book by Kilpatrick; its qualities were unsurpassed: in 122 pages there was a summary of the past, the present, and the future of a renewed education at the service of social change. In it, it was much easier to locate the pedagogical ideas of Dewey than in any other handbook. Kilptarick's writing is simple and his arguments are well schematized.[61]

Formatting Dewey as the thinker of educational renewal, relieved from his own intellectual trajectory and his religious pilgrimages, Teixeira was able to provide a non-traumatic solution to the theological enigma found in Dewey's writings, at once pragmatist and puritan, on which he based his pedagogical utopia: postulating the future of the intelligent experience against the increasing controls by the society of matter based on the idyllic image of the community of men. One should think that Teixeira's finding the spiritual tranquility in Dewey has occurred at the very point when a theology extracted from the heart of modernity has entered the emptiness left by that Catholic theology in which Teixeira had grown up in and lived with in Brazil.

From Dewey, Anísio Teixeira presented a redeeming utopia that resulted from a peculiar synthesis between Puritan theology and science; he glimpsed an urban and industrial community as a group of artisans around sophisticated machinery—No trustee, no trade union, no power to coercively interfere with that contact. On the educative guidance to direct that experience lived by the artisan, stimulating them to draw from within the art they mastered and which would be, at the same time, contained and overcome in the machine by other arts, and by other arts which, with non-traumatic ruptures, would be made science and technology. For those who lived them and made them, the art contained in them is known; for those who own the property of machines these arts are no longer of interest and time is pressing; for those who bear the art of lived life only, science and technology appear as the denial of all arts.[62]

Then, this would be the educative task for each child coming to life, for each teacher entering a classroom; and for each immigrant arriving in the United States, for each Eastern person that became a Western person. For each barbarian that became civilized. From within each of them their own controls were to be extracted, the very processes of psychophysical adaptation and the very mental adjustments to an ambience in which the outcome of the experiences of mankind, selected by intelligence, would be deposited.[63]

Through peculiar mental operations, Anísio Teixeira fused Henry Ford and John Dewey; he produced a synthesis between Fordism and the pedagogy based on intelligent experience. John Dewey—Weltanschauung, Henry Ford—organizer; as Christ and Saint Paul.[64] In 1934, Anísio Teixeira published *Em marcha para a democracia* (Marching Toward Democracy), of little circulation, in which that synthesis is patently expressed by his exposition on the "democratic [magnitudes] of North-American industrialism."[65]

This exacerbation of the idyllic image of the United States induced a colleague who was in New York in 1935 to charge from Teixeira, in a letter, more realism, since the social and educational conditions in the United States were not flourishing.[66] Teixeira then replied: "Dewey, Kilpatrick, Counts (. . .) are absolutely revolutionary in the heart of America . . . It's the very border thinking, which, to be sure, was the thinking that absorbed me in the period I was in the United States."[67]

This is not My Kingdom, But He is My Prophet

The Catholic clergy and intelligentsia never forgave Anísio Teixeira for joining the "Deweyan creed," which they took as a betrayal of Catholicism.

Even if the ex-disciple attempted to erase the traces of Dewey's trajectory and conceal his political and religious affiliations, the Catholic intelligentsia of the 1920s and 1930s could see them clearly. These men were the fruit of the aggiornamento conducted by the Catholic Church and they also had been educated in the sphere of theological and social solutions similar to those provided by Jacques Maritain. That Catholic intellectuality also pursued spiritual solutions in the heart of modernity.

Anísio Teixeira apparently did not manage to decipher this peculiar enigma: in the form he constructed of himself and Dewey, the past he had once denied still survived.

In the 1930s, Teixeira was accused of being affiliated with the Communist Party and of inciting revolution through education.[68] In the second half of the 1950s he was again attacked for defending the Act of Education Guidelines, which prioritized public school and granted government funds exclusively for public schools, the decentralized organization of the teaching system and, also, the use of pedagogical standards and practices in tune with the social demands as expressed by local communities. North American education was still his reference and Dewey was his guide.

The Catholic reaction against these theses started in 1956 with a virulence that maybe had not been foreseen by Anísio Teixeira and by the network of intellectuals who called themselves the "pioneers of the new school," whose core had virtually remained unchanged since the early 1930s. In this core, Teixeira was certainly one of the most prominent and powerful intellectuals due to his penetration both in the governmental sphere and in the academic milieu.[69]

That had not been the first occasion when Anísio Teixeira had been targeted by the Catholic clergy and intellectuals; also, it was not the first opportunity in which the Catholics associated Teixeira's educational theses with John Dewey's ideas and associated both with Marxism, with Bolshevik materialism, with revolution.

It was not, therefore, a comedy of errors and Teixeira could not live that circumstance as a farce. In none of his public writings did he put himself to deny the false relations the Catholics had set up to attack him.

From the second half of the 1950s until his accidental death in 1971, Anísio Teixeira lived a process of dilacerations as a consequence of his adherence to the social and educational ideas of John Dewey: he was attacked by the conservative leaders, namely the Catholics, because he preached the reorganization of Brazilian society based on a model containing both North American and Soviet elements. Afterward, Teixeira began to be attacked by Marxist–Leninists because he preached Dewey's liberal ideology and for being favorable to the pacific transition to a Bourgeois democracy of the North American kind.

At the end of his life, Teixeira who was disappointed with the European social and cultural standards, who never agreed with the Marxist–Leninist revolutionary choice nor with the organizational patterns of the Soviet society, suffered the breaking of the North American mirror.

From 1960 to 1970, Teixeira exchanged correspondence with his old professor, George Counts. In the very first letters, he starts to manifest a melancholic tone, which his public and personal writings had not shown for decades.

In a 1966 letter, Teixeira sent his friend a message in which his pessimism and sadness seem to have grown. He says about United States: "it is sad to see the sterile war game spirit coming from America. The most revolutionary Country, the Country of hope, turned warlike . . ."[70]

To end the correspondence, there is a last letter from Teixeira to his friend in July 1970, in which his disenchantment with the world and his nostalgia of better times overwhelm.

I keep reminding myself of your "Dare the schools to create a new order?" (. . .) other things have created a new world but not a new order. But as a soothsayer you were awake to the needs of the men. Now you are being asked "to accept," acceptance is the new wisdom. So it was when God had made the world, but now that we know that men have made it, it is not easy. Anyway, it is worse than the "God's" made world and to resign, to conform, to accept, is just the opposite of wisdom . . . to come so long and so far just to repeat such old truths when there are no soothsayers but only computers asking us for "acceptance". . . . How great look our twenties! Who is now replacing Dewey, Counts, Kilpatrick?[71]

The United States disappointed Teixeira. It is not a coincidence that, in his last letter to Counts, Teixeira utilizes a bitter rhetorical device to criticize directly his friend and indirectly the silent North American mobs: "who is replacing Dewey, Counts, Kilpatrick?" He quotes the thinkers who had been his references for almost his entire adult lifespan; for Teixeira, they were the greatest North American contemporary thinkers; but now they were dead and nobody was replacing them.[72]

A negative state of affairs; Anísio Teixeira, however, seemed to fear it could be, in fact, a lengthy period of shadows whose negative effects would last for ages.

Notes

1. I must thank the following persons who directly cooperated in the making of this text: Luiz Ramires who translated it into English and Andrew B. Boyd who conducted the technical review. Prof. Thomas Popkewitz who kindly offered excellent suggestions for the development of my arguments. Prof. Odair Sass who read the manuscript several times and helped me refine my own hypotheses. To my Monday group of students with whom I had a profound discussion about the text.

2. Anísio Teixeira occupied the highest directorial positions in public education, both state and federal in Brazil. He was a counselor with UNESCO from 1946, invited by Julien Huxley. He was a normal schoolteacher and a professor at several Brazilian universities; in the early 1960s he was a visiting scholar at the Universities of Columbia, Harvard, Chicago, California, and Michigan state. In 1963, he was awarded by *Teachers College* the Medal for Relevant Services, a prize granted to foreign professors. He left a great deal of books, chapters, articles, and other sorts of publications.

3. Micheal Foucault, "O que é um autor?" in *O que é um autor?* (Lisboa: Vega, Passagens, 1992).

4. For this chapter, they were especially the diaries, the letters, interviews, personal notes, and other materials left by Anísio Teixeira which Foucault would call "writings of oneself," that is, writings that make up "the arts of oneself," which refer to the practices of oneself and the government of the self and of others. The intention is to search these writings by Teixeira for both "the narrative of oneself"—how he shows himself, how he gives himself to be seen, how he makes his face known to the other—and for his exercises of "constitution of oneself." Says Foucault, "no technique, no professional skill may be acquired without doing it; also, one cannot learn the art of living, the *tekne tou biou*, without an *askesis*, which must be understood as a taming of oneself by oneself." See Micheal Foucault, "O que é um autor?" in *O que é um autor?* (Lisboa: Vega, Passagens, 1992).

5. H. Lima, *Anísio Teixeira: pensamento e ação*, ed. Fernando de Azevedo et al. (Rio de Janeiro: Civilização Brasileira. 1960); Filho L.Vianna, *Anísio Teixeira: a polêmica da educação* (Rio de Janeiro: Nova Fronteira, 1990).

6. The two colleges in which Anísio Teixeira taught were directed by Portuguese priests of the Society of Jesus who had been expelled from Portugal when the Republic was installed; they were representatives of the most reactionary Catholicism. The very followers of such Catholic reactionary position called themselves *ultramontanes*. It must be highlighted that in the first school of his hometown, where he entered in 1916, there were Portuguese, French, German, Swiss, and Irish priests, all of the reactionary ultramontane, and they represented a shield against the influence of the missionaries from the Presbyterian Evangelical Church, who created the American School of Caetité. See also: M. G. P. Schaeffer, *Anísio Teixeira: formação e primeiras realizações* (São Paulo: Grafon's, 1988); Vianna, *Anísio Teixeira.*

7. In Brazil, by the early twentieth century, most members of the political and cultural elites came from the courses on Law, which existed in only three cities. See, Anisio Teixeira, *Correspondência ativa e passiva entre Anísio Teixeira e Luiz Gonzaga Cabral (Pe)* (Archieve, ATc 1918.01.26, Rio de Janeiro: CPDOC/FGV, 1918); Anisio Teixeira, *Correspondência ativa e passiva entre Anísio Teixeira e Deocleciano P. Teixeira (pai)* (ATc 1922.03.06, Rio de Janeiro: CPDOC/FGV, 1922).

8. Father Luiz Gonzaga Cabral (1866–1939) was considered the greatest preacher of the Iberian Peninsula. He descended from an aristocratic family from Porto; he was the provincial of the Jesuits in Campolide at the time of the Republican revolution and was the confessor of the king. He left Portugal under multiple disguises, after being prosecuted by the Portuguese masons and the Carbonary. He arrived in Bahia, in 1916, to dedicate himself to the wakening of sacerdotal and literary callings.

9. Teixeira, *Correspondência ativa e passiva*, 1918.

10. Teixeira, *Correspondência ativa e passiva*, 1918; Teixeira, *Correspondência ativa e passiva*, 1922.

11. The countries that gave Brazil more visibility in the debates among the Catholics were England, France, Germany, Italy, and Belgium.

12. Teixeira, *Correspondência ativa e passiva*, 1918.

13. *Clientelism* is a type of political relationship in which a person (the boss) protects another (the client) in exchange for support, thus establishing a bond of personal submission which, on one hand, does not depend on akin relations and, on the other, is not construed as a legal relationship.

14. Teixeira, *Correspondência ativa e passiva*, 1922.

15. Omer Buyse, *Méthodes américaines d'éducation générale et technique* (Charleroi: Etablissement Litho de Charleroi, 1908); Omar Buyse, *Méthodes américaines d'éducation générale et technique* (Paris: D. Dunod & E. Pinat, 1909, 1913).

16. In the first part of the book dedicated to elementary teaching, Buyse reserves only one topic to what he calls "Dewey's method" within the chapter about the system of manual works with a social profile. In a few pages, Buyse describes the Deweyan experience in Chicago with the laboratory school" (Omar Buyse, *Méthodes américaines*, 1908, 1913), 129–132.

17. In his first management report as the director of public instruction, submitted to the state governor, Anísio Teixeira mentions the book by Omer Buyse

several times; based on it, he praises North American education and draws attention to the need for Brazil of the same educational renewal that was taking place in that nation. The report also includes initiatives that Teixeira had already taken to update the schools from Bahia in the North American molds, such as: the purchase of equipment, introducing new disciplines such as Drawing, Geometry, and Manual Activities. He finally adds that he has arranged for the book by Buyse to be translated in order to hand it out to teachers across the state. In fact, a small part of the book was published in Portuguese in 1927, see Omer Buyse, *Método americano de ensino* [s.t.]. (Salvador: Imprensa Oficial do Estado, 1927); Anísio Teixeira, *Diário de viagem aos Estados Unidos* (Navio Pan América, 1927). Also see, Anísio Teixeira, *Relatório da Inspetoria Geral do Ensino do Estado da Bahia* (Salvador: Imprensa Oficial do Estado, 1925); Schaeffer, *Anísio Teixeira*.

18. Anísio Teixeira traveled with the archbishop of Bahia, who also helped him plan each step of the trip.

19. Teixeira, *Diário de viagem à Europa* (Lisboa: FGV/CPDDC, 1925).

20. Teixeira, *Diário de viagem à Europa*.

21. Teixeira, *Paris é um filho espiritual de Roma. Entrevista* (*A Tarde*. Salvador, November 30, 1925).

22. Anísio Teixeira, *Paris é um filho espiritual*.

23. The "new school" movement in Brazil was ample and widespread, with a remarkable presence between the 1920s and the 1940s. Its goal was to make school work rational and make pedagogical practices scientific. Until the early 1930s, the ideological and religious differences among the defenders of the "new school" were not the object of ligitation; then differences became divergent and incompatible, namely between the Catholics and those renewers who called themselves the "pioneers of the new school," in favor of a school that should be public, that is, directed and materially supported by the government, free of religious constraints, and accessible to both sexes.

24. Anísio Teixeira, "Valores proclamados e valores reais nas instituições escolares brasileiras," *Revista Brasileira de Estudos Pedagógicos* 86 (1962): 59–79.

25. As mentioned in the previous note, the litigation between the Catholics and the "pioneers of the new school" only became sharp after the early 1930s. Thus when Teixeira began to position himself on the side of the defenders of educational renewal, they got along fine with different religious and ideological creeds. It should be realized, however, that the struggles mentioned herein did not separate Catholics and Non/Anti-Catholics in opposite fields. Most of the "pioneers of the new school" dealt with the Catholic faith as a personal or family matter.

26. A. Mayer, *Dinâmica da contra-revolução na Europa (1870–1956)* (Rio de Janeiro: Paz e Terra, 1927).

27. Mauro Passos, *A pedagogia catequética e a educação na Primeira República do Brasil (1889–1930)* (Roma: Universidade Pontifícia Salesiana, 1988).

28. One should remember here the initiatives undertaken, since the release of C. Darwin's major work, in 1859, by Catholic theologians to reconcile the evolutionist theses with the principles of Catholicism, See, M. Löwy and Robert

Sayre, *Revolta e melancholia: O romantismo na contramão da modernidade,* trans. Guilherme João de Freitas (Teixeira. Petrópolis: Vozes, 1995). Even so, the revisions made by the official Catholicism in the face of the Darwinian theses were much less profound that those occurring within some of the reformed denominations, especially in the United States. See, S. E. Ahlstrom, *A Religious History of the American People* (New Haven: Yale University Press, 1972); E. L. Blumhofer, ed., *Religion, Politics, and the American Experience. Reflections on Religion and American Public Life* (Tuscaloosa: The University of Alabama Press, 2002).

29. A. Gransci, *Cadernos do cárcere,* Vol. 4, trans. Carlos Nelson Coutinho and Luiz Sergio Henriques (Rio de Janeiro: Civilização Brasileira, 2001); Löwy and Sayre, *Revolta e melancholia.*

30. Leão XIII (Papa), *Encíclica Rerum Novarum. Sobre a condição dos operários* (Rio de Janeiro: Imprensa Oficial, 1941).

31. C. R. J. Cury, *Ideologia e educação brasileira. Católicos e liberais* (São Paulo: Cortez, 1984), Gransci, *Cadernos do cárcere.*

32. In his writings, J. Maritain (1882-1973) seeks to conciliate reason and faith, taking support from both H. Bérgson and Saint Thomas Aquinas; out of this conciliation effort comes the statement of primacy both of the being and the intelligence, and of God and "integral humanism." His core assumption is that mankind has embarked on the same collective adventure, overcoming the frontiers between nations, religions, and political power and the human community is a traveling companion, and God takes part in this adventure toward the transcendental infiniteness. See, A. S. Bogaz, *Dialética do sagrado e do profano no humanismo integral de Jacques Maritain* (São Paulo: USP, 2003).

33. One of the leaders of the educational renewal in Brazil in the 1920s was A. Carneiro Leão who directly cooperated in training Anísio Teixeira in educational matters, and who introduced him to North American education. Carneiro Leão was the one who indicated Omer Buyse's book to the governor of Bahia, and at the same time he was pointed out as the person Anísio Teixeira should take advice from in educational matters. In 1927, the script of school visits as well as the contacts with educators and scholars in the United States. were under the care of Carneiro Leão. He is the author of the only two articles by a Brazilian writer published in the *Teachers College Record* (1925, 1935). See, Teixeira, *Diário de viagem,* 1927.

34. Teixeira, *Diário de viagem,* 1927.

35. Teixeira, *Diário de viagem,* 1927.

36. Teixeira, *Diário de viagem,* 1927.

37. Teixeira, *Diário de viagem,* 1927.

38. Teixeira, *Diário de viagem,* 1927.

39. Teixeira, *Diário de viagem,* 1927.

40. Teixeira, *Diário de viagem,* 1927.

41. Gransci, *Cadernos do cárcere.*

42. The concept mentioned herein of "disenchantment of the world" comes from Max Weber. See: Max Weber, *A ética protestante e o "espírito" do capitalismo, Traduzido por José Marcos Mariani de Macedo* (São Paulo: Companhia das Letras, 2004). Out of Weber's perspective, it is at all remarkable that

Anísio Teixeira had clung to Dewey's puritan pragmatism to protect himself from the "disenchantment" when, for Weber, puritanism was decisive in the Western process of "disenchantment of the world," since it is the only ethical religion which, despite "rejecting the world" and remaining connected to a notion of "redemption," adapts itself to a typically antiethical rationalism of the secularized rational routines and, at the same time, grants it with substance. This is its major paradox: it is a religion that promotes the profanation of the world. See, W. Schluchter, *Paradoxes of Modernity: Culture and Conduct in the Theory of Max Weber*, trans. Neil Solomon (Stanford, CA: Stanford University Press, 1996).

43. Anísio Teixeira, *Aspectos americanos de educação*: Relatório apresentado ao Governo do estado da Bahia pelo Director Geral de Instrucção, Comissionado em estudos na América do Norte (Salvador: Instrucção Publica do Estado da Bahia, 1928).

44. Anísio Teixeira was especially impressed with the mutual causality that Dewey had established between democracy and education. Says he: "Only a truly efficient educational organization may uphold and maintain this ambitious project of social life that American democracy is making real. This social life of full and wide participation with no barriers and no constraints, involves a permanent confidence in the *ordinary man* (. . .). This determines a perpetual social change, absolutely indefinite, which does not degenerate into chaos only because American education seeks to provide personal initiative and social adaptability in order to create a new social balance . . ." See, Teixeira, *Aspectos americanos de educação*.

45. Out of the academic documents associated with Anísio Teixeira found in Teachers College of the University of Columbia there are two disciplines on the Philosophy of Education taught by William Kilpatrick; a course named *The economic effect of education* delivered only by George Counts and eleven more, from which Counts appears as a professor among others. See, Anísio Teixeira, *Documentação acadêmica de Anísio Teixeira do período em que permaneceu em estudos no Teachers College da Universidade de Columbia* (New York: Teachers College, Columbia University, 1928–1929).

46. It should be highlighted that Anísio Teixeira did not attend any discipline taught by Edward L. Thorndike, considering that the leaders of the movement for the "new school" in Brazil were fully aware of the importance of the studies in Educational Psychology conducted by Thorndike and his team. This fact reinforces the assumption that Teixeira established relations with North American educators through the network of relations formed by the so-called educational progressivism, which at the time fought vigorously against Thorndike. See, Miriam J. Warde, "Estudantes Brasileiros no Teachers College da Universidade de Columbia: do aprendizado da comparação," *(II Congresso Brasileiro de História da Educação*, CD-Rom, 2002). Also see, Anísio Teixeira, *Documentação acadêmica*; Miriam J. Warde, *Relatos de um certo Oriente: Gustave Le Bon, John Dewey e Omer Buyse no Brasil e na Turquia (anos 20 do século XX)* (2004).

47. Anísio Teixeira, *Caderno de aula de Anísio Teixeira* (Rio de Janeiro: FGV/CPDOC, 1928–1929).

48. Anísio Teixeira, "Por que 'escola nova'?" *Boletim da Associação Bahiana de Educação* 1 (1930): 2–30; Anísio Teixeira, "A Pedagogia de Dewey (estudo introdutório)," in John Dewey, *Vida e educação* (São Paulo: Melhoramentos, 1930); Anísio Teixeira, *Em marcha para a democracia. À margem dos Estados Unidos* (Rio de Janeiro: Guanabara, 1934).

49. The argument here is that Teixeira did not take the task of interpreting what he had "found" in Dewey: a lay religiosity which leads to the distinction that Dewey himself thoughtfully made between "religion" and "religious values," See, John Dewey, "A common faith," in *The Later Works of John Dewey*, 1925–1953, Vol. 9, 1933–1934 (Carbondale: Southern Illinois University Press. The Electronic Edition, 1986). On the other hand, in order to make up the Deweys-form, Teixeira elided Dewey's education and research within Calvinism, as well as his theological dialogues with congregationalists and unitarists, suspended by the period in which Dewey found a certain quietness in Hegel (approximately the two last decades of the nineteenth century; at the end of which, Dewey distanced himself from Hegelian logic), and he also suppressed the explicit return of Dewey to religious themes, which is patent in *Common Faith*, published in 1934. See, S. C. Rockefeller, *John Dewey. Religious Faith and Democratic Humanism* (New York: Columbia University Press, 1991); B. Kuklick, *Churchmen and Philosophers: From Jonathan Edwards to John Dewey* (New Haven: Yale University Press, 1985).

50. "New Jerusalen" and "new spiritual dwelling" are metaphors for the "community" to which Dewey claimed the restoration, as it was under attack by industrialization, science, technology and urbanization. See, John Dewey, *Vida e educação*. With Introduction by Anísio Teixeira, trans. Anísio Teixeira (São Paulo: Melhoramentos, 1930); John Dewey, *Democracia e educação: breve tratado de philosophia de educação*, trans. Godofredo Rangel and Anísio Teixeira (São Paulo: Nacional, 1936); John Dewey, *Liberdade e cultura*. With Introduction by Eustáquio Duarte, trans. Eustáquio Duarte, ([n.p.], Revista Branca, 1953/1970); E. D. Baltzell, *The Search for Community in Modern America* (New York: Harper & Row 1968); R. J. Bernstein, "Community in the Pragmatic Tradition," in *The Revival of Pragmatism: New Essays on Social Thought, Law, and Culture*, ed. Morris Dickstein (Durham: Duke University Press, 1998). Teixeira was definitely conquered by Dewey's idea of "community," as he understood that this was the effective way around which the North American society organized itself and should be adopted in Brazil; see Anísio Teixeira, A, *Em marcha para a democracia*. Dewey's debt for the idea of "community" taken from the renewed Calvinism, namely the theologian Jonathan Edwards deserves deeper exploration; Kuklick, *Churchmen and philosophers* offers elements for that task.

51. Warde, *Estudantes Brasileiros*.

52. G. M. Marsden, *The Soul of the American University: From Protestant Establishment to Established Nonbelief* (New York: Oxford University Press, 1994); C. W. Mills, *Sociology and Pragmatism: The Higher Learning in America* (New York, Paine-Whitman Publishers, 1961); L. R. Veysey, *The Emergence of the American University* (Chicago: University of Chicago Press, 1965).

53. B. Kuklick, *The Rise of American Philosophy: Cambridge, Massachusetts, 1860–1930* (New Haven: Yale University Press, 1977); L. Menand, *The Metaphysical Club* (New York: Farrar, Straus and Giroux, 2001).

54. In 1930, Anísio Teixeira published his first version of "new school" and his first open statement on behalf of it. As he saw it, "what happened was that science was applied to human civilization. Materially, our progress is the child of inventions and the machine (. . .) Science as a fact which brought with it a new mentality. First, it determined that the new order of affairs would become dynamic instead of stable and permanent (. . .) Scientific experimentation is a limited method of advancement (. . .) The new school does intend, in turn, to base itself on these facts and on this new mentality. The educational phenomenon, in Dewey's phrase, is the reconstruction of experience, under the light of past experiences, for a better and richer control of the situation (. . .) Hence, how is school to be organized on the basis of *matters to be studied*? The only *matter* for school is life itself, guided with intelligence and discrimination, so that we can make it progressive and ascending." Curiously Teixeira informs that the basis on which the new school as he was expounding it had been "outlined by Kilpatrick and, in essence, they are the same shown by several modern school reformers" See, Anísio Teixeira, "Por que 'escola nova'?"; Teixeira, *A Pedagogia de Dewey*. In writings immediately posterior, Teixeira began to include the theme of school renewal within the issues of relations between "state, society and community"; there are clues that, at that time, he started to incorporate the issues Dewey had dealt with in *The Public and Its Problems*. See, John Dewey, J. *The Public and Its Problems. An Essay in Political Inquiry in The Later Works of John Dewey*, 1925–1953, Vol. 2, 1925–1927 (Carbondale: Southern Illinois University Press, 1986, The Electronic Edition).

55. Teixeira, *Aspectos americanos de educação.*

56. In that ambience, George Counts had obtained his Ph.D. in 1916[0]. See C. D. Jay, *The Doctoral Program of George S. Counts at the University of Chicago (1913–1916)* (Carbondale, IL Southern Illinois University, 1982). It should be highlighted that also in relation to Counts and Kilpatrick, Anísio Teixeira does not present exercises of differentiation. There are no records that he had considered the peculiar aspects of Count's criticism of North American school education, its relations with Soviet education, and its dialogue with Leninist-Marxism. See B. A. Romanish, *An Historical Analysis of the Educational Ideas and Career of George S. Counts* (Carbondale, IL The Pennsylvania state University, 1980).

57. Warde, *Estudantes Brasileiros.*

58. John Dewey, "The pantheism of Spinoza," in *The Earlier Works of John Dewey*, 1925–1953, Vol. 1, 1882–1888. John Dewey, "Kant and Philosophic Method," in *The Earlier Works of John Dewey*, 1925–1953, Vol. 1, 1882–1888; John Dewey, "The Present Position of Logical Theory," in *The Earlier Works of John Dewey*, 1925–1953, Vol. 3, 1889–1892.

59. Miriam J. Warde, "O itinerário de formação de Lourenço Filho por descomparação," *Revista Brasileira de História da Educação* 5 (2003): 125–167.

60. Dewey, *Vida e educação*, With Introduction by Anísio Teixeira, trans. Anísio
     Teixeira; John Dewey, *Como pensamos*, trans. Godofredo Rangel. New
     translation and notes by Haydée de Camargo Campos [1956] (São Paulo:
     Nacional, 1933); John Dewey, *Democracia e educação: breve tratado de
     philosophia de educação*, trans. Godofredo Rangel and Anísio Teixeira (São
     Paulo: Nacional, 1936); John Dewey, *Liberdade e cultura*, With Introduction
     of Eustáquio Duarte, trans. Eustáquio Duarte, ([n.p.]: Revista Branca, 1953);
     John Dewey, *Natureza humana e a conduta* [without translator] (Bauru:
     Livraria Brasil. 1956); John Dewey, *A filosofia em reconstrução* trans.
     Eugênio Marcondes Rocha (São Paulo: Nacional, 1958); John Dewey, *Teoria
     da vida moral*, [s.t.] (São Paulo: Ibrasa, 1960); John Dewey, *Liberalismo,
     liberdade e cultura*, trans. Anísio Teixeira (São Paulo: Nacional, 1970); John
     Dewey, *Experiência e educação*, trans. Anísio Teixeira. São Paulo: Nacional,
     1971); John Dewey, *Experiência e natureza*, trans. Anísio Teixeira (São Paulo:
     Abril Cultural, 1974).
61. W. H. Kilpatrick, *A educação para uma civilização em mudança*, trans.
     Noemy da Silveira (São Paulo: Melhoramentos, 1933).
62. A. Feffer, *Between Head and Hand: Chicago Pragmatism and Social Reform,
     1886 to 1919*, (Pennsylvania: University of Pennsylvania, 1987); Miriam J.
     Warde, *Relatos de um certo Oriente*.
63. Miriam J. Warde, *Relatos de um certo Oriente*.
64. Gransci, *Cadernos do cárcer*.
65. Teixeira, *Em marcha para a democracia*.
66. M. B. Lourenço Filho, *Correspondência de Lourenço Filho a Anísio Teixeira*,
     ATc 1929.11.01 (Rio de Janeiro: CPDOC/FGV, 1919).
67. Anisio Teixeira, *Correspondência de Anísio Teixeira a Lourenço Filho*,
     LFc 30/31.05.15, (Rio de Janeiro: CPDOC/FGV, 1930); Miriam J. Warde,
     O itinerário de formação.
68. C. R. Cury, *Ideologia e educação brasileira*.
69. In the year 1956, a priest who was a congressman, writes to the minister of
     education to warn him about "the influences of Columbian pragmatism, of
     dialectical materialism" professed by Anísio Teixeira. The congressman priest
     also says that in the criticisms against the private schools in general and the
     Catholic ones in specific, Teixeira was "the revolutionary of such hidden
     campaign (. . .) inspired by John Dewey's pragmatist philosophism (*sic*) which,
     since 1932 has been the rear guard against the apostolate of the Catholic
     Church in the field of teaching." He continues: "Dr Anísio, in addition to the
     mediate danger to the education of our youth, by the preaching of a harmful
     philosophism (*sic*) by the well-known professor from the University of
     Columbia, recommended and applauded in Russia, creates a permanent
     ambience of pedagogical confusion . . . ." The congressman priest was con-
     vinced that Anísio Teixeira was "an authentic Marxist intellectual," who
     intended to implement the teachings of John Dewey into Brazilian universities
     and schools. This "rigid materialism, connected to sociological evolution,
     coincides with the major themes of historical materialism." See, Ester Buffa,

Ideologias em conflito: escola pública e escola privada (São Paulo: Cortez & Moraes, 1979).

70. Anísio Teixeira, *Correspondência ativa e passiva entre Anísio Teixeira e George S. Counts*, ATc 1960.07.01 (Rio de Janeiro: CPDOC/FGV, 1960–1971).

71. Teixeira, *Correspondência ativa e passiva entre.*

72. Teixeira, *Correspondência ativa e passiva entre.*

10

The Appropriation of Dewey's Pedagogy in Colombia as a Cultural Event

Javier Sáenz-Obregón

Overview

The chapter focuses on the discursive, institutional, and public policy appropriations of Dewey's pedagogy in Colombia in the first half of the last century. I analyze the cultural factors that explain its selective appropriation during this period. I also examine the national, political, and pedagogical context that, till 1934, tended to exclude Deweyan pedagogy, while appropriating enthusiastically the concepts and prescriptions of the European *New School* or *Active Pedagogy* movement. I also describe the power/knowledge transformations that, between 1934 and 1946 led to the appropriation of the social and political dimension of Dewey's pedagogy by teacher-training institutions and government policy. The nature of national, political, and cultural resistance to Deweyan pedagogy is analyzed in some detail, such as the power of Catholic dogma in the country, and the widely-held belief on the *degeneration* of the national race. I argue that the dualist structure of Catholic thought was particularly adverse to Deweyan pedagogy. I conclude that Dewey's tenuous welcome as a *foreign guest*, have gone hand in hand with brief periods of consolidation of democratic practices in the country, and have been more decisive when *faith* in democracy has been intensified.

A Hostile Culture: Versions of Modernity

In the first half of the last century, *pedagogy*[1] was viewed in the country as the privileged practice for the attainment of the desired moral, political, and economic ends of the nation by intellectuals and political actors with different educational agendas and versions of modernity. In 1923 Agustín Nieto Caballero, one of the most influential educational reformists, put it succinctly: "whatever the teacher is, that will be the nation."[2] This faith placed in schools for the attainment of "modern progress" turned pedagogy into a field of battle over the teacher's and the pupil's soul. This was not always an open battle: at times philosophical and political differences were underplayed in an effort to create a national consensus regarding educational policy and pedagogy. The idea being that in a country prone to political violence and with a people conceived as overpassionate, the sphere of education had to be protected from conflict. In other periods, this battle visibly pitted the Church against the state, Catholic against "modern" intellectuals, and the two dominant parties of the country: the Conservative and the Liberal Party.

Until the mid-1930s, the hegemonic discourse on "modernity" in Colombia, produced a number of what we have termed[3] *filters of appropriation* that effected a strategic cut in available knowledges and discourses produced in the United States and some European countries, that were linked to some of the most salient features of the national culture of the time: the deep divisions between the elite and the mass of the population; the great political and symbolic power of the national Catholic Church, one of the most conservative of the continent; and the fresh memories of political violence, the *Thousand Day War* (1899–1902) between the Liberal and Conservative Parties. As López de la Roche[4] has pointed out, given the extremely close relation in the country between religion and partisan politics, political controversy acquired a strongly sectarian and intolerant character. Since the middle of the nineteenth century, religion had become the major dividing line between liberals and conservatives, and was used by the two parties for electoral purposes.

Until the mid-1930s, there were three major filters that operated in the strategic selection of "modern" thought and pedagogical conceptions and practices. In the first place, a deep-seated mistrust on the part of the political, economic, and academic elite toward the poor population, which was considered to be sick and degenerate. The poor masses were viewed as a passionate, primitive, animalistic, and violent social body, riddled with physical, moral, and intellectual abnormalities. This distrust toward the poor, which constituted the great majority of the population and were the almost exclusive target of public education,[5] privileged discursive appropriations that explained their "tragic" situation as well as medical and pedagogical prescriptions for their normalization and regeneration, and for the "defense of the race."[6]

In the second place, there was a distrust toward those dimensions of the self, such as desires, emotions, and the imagination, considered to be contrary to morality, social order, and economic progress. During this period, educational reformists shared a *civilizing*[7] purpose: a project for the rationalization and modernization of popular culture that excluded those dimensions of subjectivity considered dangerous both for ecclesiastical authority and for the authority of scientific reason. Following Foucault's[8] definition, they privileged teaching and examination practices of *individualization*, through which the pupil was separated from the group so as to objectify and act upon her/him as a separate unit, as well as *dividing* practices through which pupils were led to perceive as necessary to be pitted against themselves, to suppress or control certain dimensions of their subjectivity. The effect of this filter was the design of educational purposes and programs for the formation of individuals with enough autonomy and initiative to be *productive* in the creation of economic prosperity and social order, but without opening the Pandora Box of those dimensions of the self deemed to be "irrational."

Finally, there was a third filter, which colored, as it were, the other two: the filter of ecclesiastical censorship that excluded theories, concepts, and practices contrary to the dogmas of the Catholic Church. It was a filter that beyond the defense and rejection of certain ideas, functioned as an archaeological stratum of Christian dualism—often invisible and not always linked explicitly to Catholic discourse—that was specially antagonistic to Dewey's anti-dualist philosophical and pedagogical project: his project to do away with the dualism mind/body, sensation/thought/action, interest/effort, child/curriculum individual/society, instruction/formation, morality/knowledge, means/ ends, fact/value.[9]

These filters of inclusion/exclusion made invisible the specificity of Dewey's pedagogy,[10] a situation that would only change through the relatively radical, political, and cultural transformations that took place between 1935 and 1946. Before this period, Dewey was seldom a single guest. Although Pragmatism was singled out by conservative intellectuals as inimical to Catholic dogma, Dewey was generally grouped in a section of the *traveling library* that tended to be read as if written by a single author: the section of "modern," "new," or "active" pedagogy. Intellectuals, public officials, and teachers who participated in the different factions of the pressing national debate over the purposes of education and the means to attain them, tended to overlook the differences between "modern" pedagogues. Thus, Dewey and his fellow pragmatist William James[11] tended to be grouped with the most visible representatives of the European *New School* or *Active Pedagogy* movement[12]—Montessori, Decroly, Claparéde, Ferriére—with Lay and Meumann, the German "experimental" pedagogues, and even, at times, with such nineteenth-century thinkers and pedagogues as Spencer, Pestalozzi, Froebel, and Herbart.[13]

The visibility/invisibility of Dewey's pedagogy in Colombia in the first half of the century was inversely related to that of the Belgian medical doctor and pedagogue Ovide Decroly, the most important methodizer of the ideas of active pedagogy, who visited the country in 1925. In the 1920s and early 1930s, Decroly's pedagogy was dominant in reformist discourse and educational reforms; a hegemony that was displaced by Deweyan pedagogy between 1935 and 1946. Decroly's detailed formalization of the curriculum and teaching methods and his racist and conservative social and political conceptions were openly welcomed in a period when national reformist discourse focused on the technical modernization of the country and on the healing of a "degenerate" race without openly antagonizing a reactionary Church. Within a politically conciliatory context this had the effect of blurring the radical differences between Dewey's and Decroly's pedagogy. First Decroly's and later Dewey's pedagogy were used in the country to support two radically different conceptions of modernity: one that viewed it as an increase in the efficient use of productive forces on the part of the poor population through the technical modernization of education and labor, while maintaining the "natural" hierarchical order of society; and another that viewed modernity as a process of secularization of the population, and its appropriation of the ideals, practices, and institutional arrangements of egalitarian democracy. In the first version, an ideological and symbolical continuity operated between Church dogma and practices of examination in schools and those for the "defense" of the race; while the second version of modernity was radically opposed by conservative/Catholic intellectuals and politicians.

Only in the mid-1930s, and this more implicitly than openly, did educational discourse distinguish between the racist foundation of Decroly's pedagogy, as well as the hierarchical, normalizing, and socially adaptive ends he proposed for education, and Dewey's political progressivism and conception of democracy as the ideal way of life.

Decroly's political conservatism was specially visible in the conferences he gave during his visit to Colombia in 1925. In these conferences[14] he presented his views on the need to institutionalize a *defensive school* to combat the dangers of heredity and the environment, as well as a school tailored to the biological interests of children, through a set of hygiene and individualizing practices of *examination*, which included medical and psychological testing, and the hierarchical classification and selection of pupils.[15] He presented a conception in which social and natural adaptation was both the means and the end of education. Decroly believed that social order required that each individual occupy his corresponding place in society, according to the "imperatives" of his own nature. This individual nature defined the precise and verifiable place that each human being occupied in the evolutionary scale. Decroly held that the weight of heredity is largely unmodifiable. While in Colombia, Decroly also emphasized his views on geographical determinism, linking climatic conditions to the evolutionary possibilities of the child and of society; in his words: "the great cultures have always flowered in the plains or in the high plateaus."[16] He also underlined a view of the hierarchical order established by the workings of the natural law; that is, a *natural selection* between a necessarily small elite and the mass of the population, arguing that for the latter, a practical education based on manual activities was sufficient.

Five main cultural events can be discerned in relation to the appropriation and opposition to Dewey's pedagogy in Colombia in the last century. First, between 1903 and 1930 there was a relative invisibility of Deweyan pedagogy within the debates generated by the reformist pedagogical movement promoted by "modern" intellectuals, and opposition by conservative reformists to Pragmatism. Second, institutional appropriations of active pedagogy in the 1920s and 1930s, based on practices of examination and the precepts of the *defensive school*, and one of the earliest institutional appropriations of Dewey's pedagogy in Latin America: that of the private reformist school for the elite— the *Gimnasio Moderno*—founded in 1914. Third, a more open and political appropriation of Dewey's pedagogical concepts and precepts in the 1930s and first half of the 1940s, in the context of partisan liberal governments. Fourth, the educational and pedagogical counter-reform initiated in the late 1940s under conservative governments that attacked and effectively erased from public memory any vestiges of Deweyan pedagogy. Finally, in the 1990s, some elements of Deweyan pedagogy would become visible again.

Pedagogical Debates and Projects of Reform: 1903–1930

Until the mid-1930s, there were few explicit references to Deweyan pedagogy within the national debate on educational reform. In the country, Dewey had yet to become a singular pedagogue. The main object of reformist discourse in this period of moderate Conservative Party governments, was the generic *active* or *new* pedagogy that privileged appropriations from Decroly's pedagogical conceptions and precepts in the process of local and national educational reform.

In the second decade of the last century, the first reform proposals through the appropriation of the ideas and practices of European active pedagogy were made in the context of a public educational system dominated by a hybrid organization of elements that I term the *official confessional pedagogy* of the country.[17] This pedagogy was the result of the selective appropriation of Pestalozzi's *objective lessons,*[18] the authoritarian disciplinary and moral educational precepts of the Christian Brothers' Catholic congregation,—based on *La Conduite*, the pedagogical treatise of their French founder, Jean Baptiste de La Salle—elements derived from classical or "rational" psychology and neothomist philosophy, and the dogmatic orientation of the Catholic Church.

In 1903/1904, the government passed an educational law of curriculum reform for primary and secondary schools[19] that defined the central purpose of education as the formation of a productive individual that was hard-working, practical, healthy, and with initiative: an individual conceived in the image of the German and Anglo-Saxon "races," thought to be the main bearers of these virtues. The law constituted a project of individualization for the production of the type of individuals deemed necessary for a modern organization of labor and production. Despite these modernizing elements, the law also gave a central place to Catholic moral and religious education that was to be chiefly directed to the rejection of "modern" and "revolutionary" errors, which for the Church included Protestantism, as well as Pragmatism and other "materialist" systems of thought. The law established the field of legitimacy and possibility for an educational system subjected to Church surveillance: the acceptance of modern methods of instruction and sociobiological and economic purposes, and the exclusion and rejection of non-Catholic moral practices and purposes.

With the creation in 1915 of the *Cultura* journal by a group of relatively progressive liberal intellectuals, a systematic diffusion of active pedagogy began through a series of articles written by national reformists, which would later be strengthened by several new publications—some of them relatively massive and published by the national and some local governments—as well as by a number of pedagogical, medical, psychological, and hygienic treatises written by national partisans of active pedagogy. In very schematic terms, until the mid-1930s, intellectuals, pedagogues, state officials, and politicians that proposed educational reforms based on active pedagogy, shared a set of conceptions that help to explain the relative invisibility of Deweyan pedagogy in public debate.

In the first place, the Progressive Liberals shared the same type of dualist "either-or" rationality in the field of pedagogy that Dewey[20] criticized in the *Progressive Education Movement* of his own country. This dualism was particularly evident in the dichotomy established between "tradition" and "modernity." Following the Enlightenment myth of the greater value and truth of what was "new,"[21] these intellectuals critiqued anything that was "old," under the strategic term of the "traditional." One of the main targets of their "modern" critique was the "traditional school." This "traditional school" constituted a fabricated image rather than a careful scrutiny of the existing pedagogical conceptions and practices in the country, and was attacked as the depositary of all the "vices" and negations *vis a vis* the ideals of active pedagogy. Each principle of active pedagogy was paired with a negative characteristic attributed to the traditional one, thus: child/teacher, activity/passivity, understanding/memory, freedom/ authoritarianism, interest/effort, innovation/tradition, and observation/verbalism.

In the second place, they shared a reductive vision of modernity, founded on biological and positivist conceptions. Their "modern" discourse was based, primarily, on sociobiological speculation and strategic appropriations of evolutionary theories and on an empiricist epistemology. Modernity tended to be reduced to what was practical, measurable, visible, useful and "natural." Childhood, a privileged object of knowledge for the "modern" disciplines appropriated in reform proposals and teacher-training institutions in this period—"scientific" administration, hygiene, and experimental medicine, physiology, and psychology—had a central place in this configuration and was viewed as the paradigmatical symbol of the future and as the "seed" of the individuals that would modernize the country. For this "modern" discourse, the central purposes of life and education were struggle for survival, adaptation, natural and social selection, and progress, understood as economic growth and the "regeneration" of the national race.

In the third place, in general terms, these reformists had a common view of some central pedagogical concepts and of the type of practices that should be introduced in schools, which in some key issues were radically different to those of Deweyan pedagogy. One of the major differences had to do with their conception of the possibilities and purposes of education. On the one hand, national reformers focused on the limits rather than the possibilities of education, emphasizing a series of factors—individual heredity, geographical and racial determinism, the need to preserve social order and to regenerate the race—that fixed within rather narrow limits what was seen as possible and desirable. On the other hand, within reformist discourse, means and ends were sharply separated. Following conceptions such as those of the Swiss active pedagogue Claparéde,[22] reformist pedagogy during this period conceived its methods and techniques as "neutral" means that were efficient and effective for the attainment of *any* end.

There was also a fundamental discontinuity with Deweyan pedagogy in their conception of what constituted a pedagogical *experiment*. While for Dewey, experimentation was a form of experience with open results through which human beings transformed their context and reconstituted their selves,[23] the conception of experimentation of national reformers was a direct appropriation of the notion of experiment in "scientific" psychology: the introduction by expert knowledge of predefined modifications in a controlled context—that is, in *efficient* schools that ideally would eliminate any activity foreign to the experiment—and the measurement of their effects on pupils through specialized techniques: anthropometric, physiological, medical, and psychological tests.[24]

A final difference between reformist discourse and Deweyan pedagogy was regarding practices of instruction and formation. In reformist discourse in Colombia during this period, prescriptions for instruction and formation were separated both explicitly and implicitly, and this was an effect of a radical differentiation between morality and knowledge. Under the strict surveillance of the Catholic Church, which believed it could hold on to moral sovereignty over schools if religious instruction and Catholic ends were preserved, it was licit to propose the modernization of methods of instruction, but reformers were evidently cautious in relation to alternatives for the moral formation of pupils. Whereas for Dewey all human practices were moral, the dogmatic defense by the Church of morality as its sacred property, together with the concerns of national reformers as to the "moral weakness" of the population, created a sharp division between prescriptions for instruction and those for formation. In this sense, Catholic practices of moral formation, and those of the *examining* and *defensive* school privileged by reformist discourse during this period, shared a common mistrust toward the national population, whether this mistrust was expressed in terms of "sinfulness" or sickness/degeneracy.

This common mistrust meant that moral education, in both cases, had an authoritarian character. It was an exercise of open power on the part of subjects of expert knowledge—theological, psychological, or medical—on pupils, in order to produce virtuous, healthy, and "normal" individuals. The main effect of this instruction/formation divide was that while, until the mid-1930s, educational reformers proposed unequivocally the adoption of relatively liberating methods for the development of knowledge—learning through action, the integration of subject matter, the development of learning habits rather than the accumulation of facts, and the introduction in schools of games, manual work, and field trips—in terms of formation they emphasized what in the documents of the time is referred to as "moral orthopedics": the diagnosis, control, and surveillance of pupils' morality, and their moral correction and "straightening."

Concurrently with the enthusiastic diffusion of European active pedagogy in the 1920s and 1930s, conservative educational reformists singled out Pragmatism as contrary to Catholic dogma. In his psychological treatise for teachers, for example, Sieber,[25] the director of one of the most important teacher-training institutions of the country attacked the pragmatist notion of the physical basis of emotions. While the conservative reformist pedagogue Jiménez López,[26] had this to say on the effects of Pragmatism in education: "pragmatist doctrine proscribes all knowledge that cannot be applied directly to the needs of life, as was to be predicted . . . this led to a reduction of the spiritual horizons of youth, that seriously weakened the highest and most beautiful attributes of human personality." For Jiménez López, the greatest sin of Pragmatism was that by prescribing pedagogical practices for the unconscious formation of habits of action, it undermined the idea of the autonomy of willpower as a separate faculty of the soul and center of the spiritual life, thereby putting into question a central tenant of Christianity: the dogma of free will, without which individuals could no longer be judged as sinful by teachers, priests, the Church, and God. For Jiménez López,[27] "a morality founded on the force of habit has no merit whatsoever . . . one cannot talk about virtue where effort and the triumphant action of will-power are missing."

Institutional Appropriations of *Active* and Deweyan Pedagogy: 1914–1930

From the early 1920s, under the orientation of medical knowledge, practices of *examination* and active teaching methods were introduced in some schools for "abnormals" and young delinquents, as well as in regular schools. It was medicine that introduced the image of the omnipotence of methods based on "scientific" precepts that would dominate national discourses of pedagogical reform until the mid-1930s. The central purposes of these institutions was the detection and correction of the physical and psychological abnormalities of pupils; the "economy" of pupils' efforts by reducing the fatigue produced by overburdened study plans, and the increase in academic productivity and their future "social efficiency." These institutions, through a complex system for the production of information on pupils—medical, psychological, and anthropometric examination; the study of their hereditary history; and the observation of their personality and moral traits—sought to predict and control pupils' performance.

Also in the 1920s, the most important local educational reform, in the state of Boyacá, put into practice the precepts of the *defensive school*, together with the introduction of Decroly's teaching methods. This reform tendency was founded on a sociobiological discourse that appropriated evolutionary theories, as well as the notion of racial degeneration that would dominate national, social, and educational debate until the mid-1930s. In the 1918 *Debate on the degeneration of the national race*, racial degeneration had been defined by Jiménez López[28] as a regression in the vital and productive capacities of the national race in relation to the original European, African, and Indian races. The national race would constitute a "sickly deviation of a primitive type," caused by external influences—climate, nutrition, education—and hereditary factors. According to Jiménez López, degenerate individuals, which would comprise the majority of the poor population, were unable to procreate normal children. The Boyacá reform emphasized the improvement in pupils' nutrition and of the hygienic conditions of schools, and the outreach of schools for the control of the biological and moral life of the rural population: procreation, birth, nutrition, infant education in the family, and the prevention of alcoholism and disease.

The *Gimnasio Moderno*, a private school for the elite founded in 1914, was the only institution that unequivocally appropriated elements of Dewey's pedagogy in the first three decades of the last century. Its founder, Agustín Nieto Caballero, had come into direct contact with the ideas of active pedagogy in Europe and the United States, including Dewey's, at Teachers' College. The pedagogy of the Gimnasio constituted a sharp discontinuity with the dominant reformist movement, primarily through the appropriation of Dewey's political conceptions and his vision of schools as social, rather than examining or defensive institutions. Its central purpose was the contribution to the development of democracy, social welfare, and solidarity. The school emphasized pupils' social responsibility and their participation in the resolution of the social problems of the country, and following Dewey's conception that moral/social education could not be a specific instructional subject but the result of the ways schools were organized, for Nieto: "the foundations of democracy are not in school-texts . . . the feeling of social responsibility can never be created within a disciplinary régime that stifles the personality. In order to have democracy, we must educate within democratic principles."[29]

Against the individualizing practices prevalent in the reformist movement, the school introduced collective practices of cooperation, as well as the Deweyan notion that teachers were to develop knowledge of pupils through "sympathetic" understanding. The Gimnasio was a hybrid pedagogy: a synthesis of Decroly's teaching methods and the social conceptions and political purposes of Dewey's pedagogy. While using the general framework of Decroly's teaching system based on *Interest Centers*, Nieto socialized the Belgian's conception of childhood *interest*. Rather than predefined by a universal evolutionary law, Nieto considered interests in a more Deweyan fashion, as social interests of cooperation and solidarity. But the Gimnasio was far from being a Deweyan school: it was only for boys, it explicitly rejected a participatory school government, and included Catholic religious education that was not compulsory for pupils of other creeds. There were two major reasons for this: the initial opposition of the Church hierarchy to the school, and Nieto's view that the Catholic religion, followed by the great majority of the population, constituted an important unifying factor for a nation prone to social and political violence.

From Biology to Society: Dewey and the Politization of Education, 1930–1946

With the arrival to power of the Liberal Party in 1930, a process began that temporarily displaced the hegemony of Catholic, biological, psychological, and medical conceptions and school practices; and that, for the first time in the century, was dominated by a secular and democratic vision of education. Within this process the divide I have described between instruction and formation, as well as that between the means and the ends of education was significantly weakened, and Deweyan pedagogy was privileged in the state's discourse and educational reforms. Through a strategic focus on the democratic organization, the collective nature, and the political purposes of schools, liberal governments effectively socialized and politicized the educational system. This process can be separated into two distinct periods. The first, from 1930 to 1934, in the context of a moderate and politically conciliatory government that gave participation to conservatives and kept harmonious relations with the Church, in which the new political discourse and Deweyan practices acquired greater visibility but coexisted with previous discourses and practices. In the second period, from 1935 to 1946, characterized by openly partisan, and at times dogmatically sectarian and anticlerical governments, in which the political dimension of Deweyan pedagogy became hegemonic. In this period, education became, as it had been in the nineteenth century, the principal sphere of political conflict, in which the most radical elements within the two parties battled each other, and in which the first signs appeared of the political violence that would wreck havoc in the late 1940s and 1950s, when some 300,000 people were killed in the context of Liberal–Conservative violence.

Under the leadership of Nieto, now a high-ranking official at the Ministry of Education, new programs for primary schools were introduced experimentally in some regions of the country in 1933. These programs, that became mandatory for all schools in 1935, were virtually the same as the pedagogical system that Nieto had designed in the *Gimnasio Moderno*. Nieto, faithful in this respect to Dewey's radical egalitarianism, saw no reason to modify in any substantial way the pedagogical practices implemented with an elite student population, which were now to be introduced for the overwhelmingly poor population of the public education system.

All the major transformations introduced by the government's programs pointed in the direction of a "socialization" of pedagogical

practice that distanced it from its previous racial and examining referents, through the appropriation of Deweyan concepts and prescriptions. In the first place, the exclusion of religious instruction in schools
and of medical and psychological examination practices; in the second
place, modifications in the content of Decroly's *Interest Centers*, which
had been adopted as the general framework of the new study plans.
Decroly's system emphasized the child's relations with the natural
world and the "natural needs of humanity," and organized the curriculum around these: the need for nourishment, for shelter, of defense
against natural and social dangers and, to a lesser degree, the need to
work in solidarity.[30] In the new national programs, social rather than
natural interests, became the organizing factor of subject matter and
teaching. Teachers' and pupils' vision and minds were directed to
social and national phenomena and problems. Social relations were
the articulating core of all content. Each year a different "interest"
was developed: life in the family and school; life in the community;
life in society, and geographical, economic, and social aspects of the
country.[31] In the third place, the program's prescriptions on formation,
school discipline, and the organization of pupils' activities also constituted a direct appropriation of Deweyan pedagogy. In clear reference to
Catholic and classical practices of formation, the programs considered
that democratic virtues could not be formed through the simple teaching of moral or civic precepts; moral/social habits were seen as the result
of a school environment in which democratic values were practiced in
everyday affairs. In discontinuity with previous discourse and practices
that viewed education as the formation of *individuals*, the programs
viewed schools as "small societies" and prescribed a discipline founded
on collective work. Discipline was to be a result of activities chosen by
pupils in whose direction they participated. As in the *Gimnasio*, the
ends of education were social and political: the purpose of schools was
no longer seen as the defense of the race or the "natural" selection of
children, but as the formation of citizens for a democratic life.

From 1935 onward, in quite a decisive manner, the educational
purpose of democratizing and secularizing national culture and the
relations between society and the state, became paramount in educational discourse, public policy, teacher-training institutions, and schools.
The formation of citizens through democratic practices in schools,
now viewed as laboratories of democracy, became a pressing concern
for the liberal government in power; a concern that was related to a
new vision that, despite the introduction of active pedagogy reforms,
schools and teacher-training institutions in the country remained

authoritarian organizations based on "perverse" Catholic and "military" systems of discipline. This new concern was related not only to the furthering of partisan goals directed to the conversion of the poor population to liberal ideals, thereby loosening its historical political ties with the Church, it was also linked to the urgent defense of the country's democratic form of government against international Fascism and Nazism and its local followers in the Conservative Party. In this new political context, and for the first time in the century, state officials dared to criticize openly the authoritarian character of educational institutions run by Catholic congregations. The annual reports of the now militantly liberal directors of public education at the local level, underlined in no uncertain terms the democratic purposes of the new liberal school that had to be created. In 1937, the director of public education of Boyacá, one of the most conservative states in the country, synthesized the liberal government's task as that of "providing individuals with mental and economic freedom, in order to make possible their self-determination, so that the impositions which arise from privilege and social artifice do no affect them."[32] These reports were especially critical of the "public" education provided by Catholic congregations in Native Indian communities, through a contract with the state that funded them, which in some regions was still operating in the 1980s. The conclusion of the local director of education in his report of an official visit to an Indian region in the state of Cauca[33] was that these contracts, established in the first decades of the century, had to be reformed for the education imparted by the missionaries was a "total failure" and contrary to the government's educational policies. In his view, the missionaries had no knowledge whatsoever of national educational programs, and the education they imparted made use of religious dogma in order to subject the willpower of Indian pupils and was centered on the memorization of prayers and religious texts, and rather than developing the pupils' personality, had the effect of destroying their happiness and freedom.

A new *filter of appropriation* of available knowledges rapidly replaced the previous ones: their usefulness in relation to the new social and political ends of the party in power. Public education policy and programs had a clear target: the integration of the rural population to a modern and democratic culture. In 1935, in its political platform, the ruling Liberal Party had offered the establishment of lay public schools, and the rejection of "reactionary dogmatisms,"[34] and in 1936 a concstitutional reform introduced for the first time in the country the principle of freedom of teaching and religious conscience in public education. In this period, schools began to be seen as institutions for the creation of new democratic, political, and economic habits, both in pupils and in the rural population in general. The teachers' gaze was redirected from the biological laws of the evolution/ degeneration of childhood and the national race, to the understanding of the culture and living conditions of the rural poor. With the appropriation by teacher-training institutions, public officials and rural education programs of *social knowledges* such as sociology and ethnology, the interdisciplinary relations of pedagogy were multiplied, and new objects of discourse and knowledge appeared—such as culture, society, politics, and nationality—that displaced those of the experimental sciences. The notion of *race* was displaced by the conception of *culture*, and sociobiological speculation began to be replaced by sociological and anthropological studies on the characteristics of national culture that produced a new and positive image of the Indian population.

These new referents of power/knowledge made possible, for the first time, an explicit critique of the conservative purposes and effects of the dominant tendency of active pedagogy appropriated previously in the country. This included a critique of the racist conceptions that had dominated in the country since the beginning of the century. A critique that was founded on a new trust in the potentialities of the "people"—*pueblo* in Spanish, that named, above all, the poor population—and the rejection, in words of Jorge Eliécer Gaitán, minister of education and the most important popular leftist political leader in the country,[35] of the "pessimism" of the defenders of an "imaginary collective decadence."[36] It also encompassed a critique of the practices of "natural selection" in schools. In this regard, in his educational treatise, based almost completely on Deweyan ideas, a national pedagogue stated: "the purpose of education is not the ingenious adaptation of the individual so that he will occupy a pre-established place in society, but rather that of enabling him to create a place for himself."[37]

These transformations went hand in hand with a very visible shift in the pedagogical foundation of state discourse and policy: a shift from Decroly and the generic referents to active pedagogy to the political dimension of Dewey's pedagogy, which nevertheless still tended to exclude the more complex ethical and epistemological foundations of Deweyan thought. This appropriation was only possible under hegemonic Liberal governments that unlike those that had preceded them, were very not overly concerned about ecclesiastical opposition. Decroly's conservative pedagogy lost its appeal in the context of the political ends of the state, while, the Deweyan conception of the function of schools in the democratization of social relations, became the new cornerstone of educational and pedagogical reform.

The new political discourse of the state and the secularization of education, together with the appropriation of Dewey's democratic philosophy, clearly challenged the foundations and purposes of the Church and Catholic dogma, and came increasingly under the fierce attack of priests and conservative intellectuals. Educational reform was attacked as atheist, materialist, and communist; as well as through a series of tactics that sought to prevent children from attending public schools. These included the deployment of priests and their exhortations during mass in order to convince parents to move their children to Catholic private schools; and the excommunication, in some regions, of those who refused. Once more, Pragmatism was singled out as one of the central enemies of Catholic ideals. In the words of the conservative pedagogue Bernal Jiménez, one of the leaders of active pedagogy reform in the 1920s and 1930s, and now Catholic crusader against liberal reforms:

> Neither ethical truth, nor juridical truth, are true as a result of their practical efficacy, as William James' pragmatism holds—the most despicable of the modern derivations of positivism—they are true because of the absolute content with which they are infused, due to their being emanations of divine law.[38]

For Jiménez López[39] the pragmatist foundation of the new reforms was evident, and, in his view, were leading to class conflict, and were supported by anti-Catholic agents and forces, such as Protestantism, as well as by exiled Jewish intellectuals, whom he considered an "undesirable ethnic element." Jewish intellectuals had been welcomed by the liberal regime and had played an important role in educational reform. For Jimenez' feverishly dogmatic and anti-Semitic mind, this unholy alliance of Pragmatism, Protestantism, and progressive Jewish intellectuals was undermining the country's "Christian civilization."

One of the main effects of the state's new social and political vision was the transformation of teacher-training institutions. Until 1934 they had remained under the symbolic and often direct control of ecclesiastical power. They had sought to form a "teacher-apostle"; a teacher formed by the moral precepts of Catholicism. The teacher was also to be an expert in the active methods of instruction, in experimental psychology, in examining and hygiene practices, and in the new conceptions of childhood created by scientific psychology and active pedagogy. By 1937, the ideal features of the "good teacher" had been transformed. She was no longer identified with Catholic virtues nor with medical and psychological knowledge, but with rather Deweyan "social" capacities, such as cooperation, social action, and *sympathy* for pupils.[40] Between 1934 and 1946 different elements of Deweyan pedagogy were introduced in teacher-training institutions. These included participatory forms of government and disciplinary regimes founded on moral self-direction and collective work. The new visibility of Deweyan pedagogy can be attested in the graduation theses presented by teachers during this period. They show that the newly formed teachers had a clearer vision of the ethical and epistemological specificity of Deweyan pedagogy than the more visible reformist leaders.[41]

Dewey, Democracy, and National Culture: Conservative Counterreform and Reemergence of Deweyan Pedagogy in the 1990s

In 1946, with the return of the Conservative Party to power, one of the most audacious periods of pedagogical reform in the country came to an abrupt halt. In the context of the beginning of a new bloody period of political violence, the conservative governments as well as liberal intellectuals, attacked the pedagogical reforms of 1935–1946 as one of the main causes for the moral and social "disorder" in the country. Public education was once again founded on an authoritarian pedagogy based on Catholic dogma that defined its central purpose as the return the country to its "traditionalist and Catholic" course, from which it had been deviated by the "corrupting materialism" of liberal governments.[42] It was a purpose that, for the conservatives, could only be attained by educating a ruling minority that would contain the passions of the poor population, viewed again as sick and prone to moral and social dissolution.[43] Liberal educational reforms were held responsible by conservatives for the outbreak of violence, while the Catholic Church, engaged in a "holy crusade" against liberalism, went as far as warning that no Catholic should vote for the Liberal Party. Liberal political leaders and intellectuals, on the other hand, blamed the reforms for their "mistakes" in the moral education of the poor population that, in the words of a former reformist intellectual, had produced "impulses of atavistic resentment" amongst the poor.[44] The poor had to be defended from democracy. Only in the early 1990s would Dewey's pedagogy be referred to again in teacher-training institutions and public debate. Once more, its political dimension was privileged, in the context of a renewed faith in democracy, following the progressive constitutional reform of 1991 that sealed a peace treaty with guerrilla organizations. Till then, a conservative political culture and the hegemony of Marxist discourse amongst opposition intellectuals acted as barriers to Deweyan pedagogy.

The destiny of Deweyan pedagogy in Colombia has gone hand in hand with the institutional possibilities for the exercise of democratic practices and with the symbolic value placed on them. The indigenization of Dewey in Colombia has had an almost exclusively political character. His pedagogy has been welcomed as an instrument for the furthering of democratic practices and institutional arrangements, and has been made invisible or openly opposed in times when democracy has been viewed as dangerous for morality, social order, and progress. If within the educational debates in the United States, Dewey clarified, modified, and contrasted his views in relation to those of his followers and opponents;[45] the precarious epistemological existence of his discourse in Colombia, served to blur its specificity and neutralize some of its radical potentialities. We are only now beginning to recognize that, in the long run, Dewey's educational project was defeated in this country by the same conservative forces of expert and *efficient* control of the teacher, pupils, and schools that were victorious in his homeland during his own lifetime.

The precarious existence of Dewey's pedagogy in Colombia can be attributed, primarily, to Catholic/conservative opposition and to the version of modernity that prevailed in national debates and in the direction of the educational system. It was a version that, with the exception of the brief democratizing period between 1935 and 1946, emphasized the technical and economic *modernization* of the population, rather than political and cultural *modernity*. Dewey's intensification of some of the more radical, political, epistemological, ethical, and aesthetic elements of modernity could not be welcomed for long in a country with a dominant culture whose distrust of the majority of its population has proven to be stronger than its faith in democracy.

Notes

1. Building upon Zuluaga's conceptualizations—O. L. Zuluaga, *Pedagogía e historia* (Bogotá: Ediciones Foro Nacional por Colombia, 1987). I define pedagogy as a *discipline that conceptualizes on, and applies and experiments with*

knowledge relative to instruction and formation in schools. Pedagogy would be, then, both a knowledge and a practice that includes concepts, precepts, and aims relative to a wide array of themes, such as study plans, teaching, learning, teachers, human nature, childhood, knowledge, morality, discipline, and examination.

2. Caballero, A. Neito, "El problema máximo," in *Escuela Activa. Selección de textos* (Bogotá: Presencia, 1923), 111.

3. J. Sáenz-Obregón, O. Saldarriaga, and A. Ospina, *Mirar la infancia: pedagogía, moral y modernidad en Colombia, 1903–1946*, Vols. 1 and 2 (Bogotá: Colciencias, Uniandes, Foro, Universidad de Antioquia, 1997).

4. F. López de la Roche, "Tradiciones de cultura política en el siglo xx," in *Modernidad y sociedad política en Colombia*, ed. M. E. Cárdenas (Bogotá: FESCOL, Foro Nacional por Colombia, IEPRI, 1993).

5. In comparison with other Latin American countries, historically there has been in Colombia a marked tendency for only the poorest sectors of the population to send their children to public primary and secondary education, while, in the first half of the last century, Catholic congregations dominated private education. In the 1920s, half of the secondary school population, which included very few poor pupils, went to private institutions.

6. This notion of defending the "national race" as a way of protecting society constituted a recontextualization of the discourse on race in Europe and the United States. On this see Michel Foucault, *Il faut défendre la société*, Cours au College de France, 1976 (Paris: Seuil/Gallimard, 1976); Stephen J. Gould, *The Mismeasure of Man* (New York: W.W. Norton, 1981).

7. On the process of "civilization" in Europe since the Middle Ages and its effect of demanding greater control on the part of the individual over his/her emotions, see Norbert Elias, *The Civilizing Process: The History of Manners and State Formation and Civilization* (Oxford: Blackwell, 1977/1994).

8. Michel Foucault, *Discipline and Punish: The Birth of the Prison* (London: Penguin, 1975/1979).

9. John Dewey, "The child and the curriculum," in *The Philosophy of John Dewey*, ed. J. J. McDermott (Chicago: University of Chicago Press, 1902/1981); John Dewey, "Interest and effort in education," in *The Middle Works, Volume 7: 1912–1914*, ed. J. A. Boydston (Carbondale and Ewardsville: Southern Illinois University Press, 1913/1979): John Dewey, *Democracy and Education: An Introduction to the Philosophy of Education* (New York: The Free Press, 1916/1997); and John Dewey, *Experience and Education* (New York: Collier Books, 1938/1962).

10. Most of Dewey's major pedagogical works or selections thereof were translated into Spanish before the 1930s, while *Experience and Education* was translated in 1939, one year after its English publication.

11. William James's *Talks to teachers on psychology and to students on some of life ideals* (1899) had been translated and published in the 1920s in a national pedagogical journal and was reprinted in 1941 by the National Ministry of Education as: William James, *Charlas pedagógicas* (Bogotá: Imprenta Nacional, 1941).

12. *Active Pedagogy* or the *New School* are terms that are used to name a rather heterogeneous reform tendency initiated in the first decades of the last century, which sought to transform existing instruction and formation practices in schools. Although primarily a European movement, many include within this tendency the *Progressive Education Movement* in the United States, which had in Dewey one of its first leaders. According to Oelkers— Jurgen Oelkers, "Break and continuity: observations on the modernization effects and traditionalization in international reform pedagogy," *Paedagogica Historica* xxxi, no. 3 (1995): 675–713—in Europe, it was first institutionalized in 1920 with the creation in London of the national *New Education Fellowship* and with its first international congress of 1921, where the *League for the Renewal of Education* was formed. By 1932, the movement had become an international phenomena, with the participation of representatives of fifty-three countries in its Nice conference. According to Adolphe Ferriére, editor of the French journal of the movement, its pioneers were Decroly, the Frenchmen Bertier and Cousinet, the Italian Lombardo-Radice, and Ferriére himself. Other pedagogues that are often named as part of this tendency are Montessori, Binet, Kerschensteiner, Claparéde, and Piaget.

13. The main historical source for the analysis of the appropriation of Deweyan pedagogy in Colombia is Sáenz-Obregón, Saldarriaga, and Ospina, *Mirar la infancia*, 1997.

14. O. Decroly, *El doctor Decroly en Colombia* (Bogotá: Ministerio de Educación Nacional, 1993).

15. Following Foucault (1975/1979, 183, 191–192) I understand examination as a practice through which individuals are governed and knowledge on them is produced. Examination objectifies and judges individuals: they are compared, hierachized, and excluded, thereby ensuring the disciplinary functions of distribution, selection, classification, normalization, and the maximum extraction of forces and time.

16. O. Decroly, *El doctor Decroly en Colombia*, 119.

17. On this, see J. Sáenz-Obregón, Saldarriaga, and Ospina, *Mirar la infancia*.

18. Appropriated by public education in the country since the mid-nineteenth century, through the American teaching manuals of H. Wilson (1870), N. Calkins (1872), J. P. Wickersham (1882), J. Baldwin (1902), and J. Johonnot (1911).

19. A. J. Uribe, *Instrucción pública* (Bogotá: Imprenta Nacional, 1927).

20. Dewey, *Experience and Education*.

21. On the idealization of *novelty* in pedagogical discourse, see J. Sáenz-Obregón, *Pedagogical Discourse and the Constitution of the Self* (unpublished doctoral thesis, London: Institute of Education, University of London, 2003).

22. E. Claparéde, *Psicología del niño y pedagogía experimental* (México: Editorial Continental, 1905/1959), 100.

23. On this, see Dewey, *Democracy and Education*.

24. This conception of the pedagogical experiment was founded on appropriations of Claparéde's (1905/1957) conceptions, as well as those of the German pedagogues,—E. Meumann, 1907/1966. *Pedagogía experimental*, 5th ed.

(Buenos Aires: Losada, 1907/1966)—and W. A. Lay, *Pedagogía experimental* (Barcelona: Labor, 1928).

25. J. Sieber, *Psicología para escuelas normales y maestros* (Tunja: Facultad Nacional de Educación, 1933).

26. Miguel Jiménez López, *La escuela y la vida* (Bogotá: Banco de la República, 1928), 115.

27. Miguel Jiménez López, "Lo inconsciente en la educación," *Cultura* 1, no. 2 (1915): 387.

28. Miguel Jiménez López, "Nuestras razas decaen: algunos signos de degeneración colectiva en Colombia y en los países similares," in *Los problemas de la raza en Colombia* (Bogotá: Biblioteca Cultura, 1928).

29. A. Nieto Caballero, *Una escuela* (Bogotá: Antares, 1966).

30. O. Decroly and G. Boon, *Iniciación general al método Decroly y ensayo de aplicación a la escuela primaria* (Buenos Aires: Losada, 1937/1939).

31. República de Colombia, *Programas de ensayo para escuelas primarias* (Bogotá: Imprenta Nacional, 1933).

32. Dirección de Educación Pública de Boyacá, *Informe del Director de Educación Pública* (Tunja: Imprenta del Departamento, 1937).

33. Dirección de Educación Pública del Cauca, *Informe del Director de Educación Pública* (Popayán: Imprenta del Departamento, 1935).

34. G. Molina, *Las ideas liberales en Colombia* (Tomo tercero. Bogotá: Tercer Mundo, 1987), 19.

35. Gaitán who was defeated in the 1946 presidential election was murdered in 1948. His death led to popular uprisings and intensified political violence throughout the country.

36. J. E. Gaitán, *La obra educativa del gobierno en 1940* (Bogotá: Ministerio de Educación Nacional, 1940), xi.

37. L. E. Pinto, *Reflexiones de un educador* (Bogotá: Editorial Kelly, 1946), 72.

38. R. Bernal Jiménez, *La educación he ahí el problema* (Bogotá: Ministerio de Educación Nacional, 1949), 281.

39. M. Jiménez López, "La actual desviación de la cultura," in Jiménez López, *Discursos y ensayos* (Tunja: Imprenta Oficial, 1948).

40. Ministerio de Educación Nacional, "La Ficha del Maestro," *Revista del maestro* 1, no. 3 (1937).

41. For an example of this, see, J. C. Feuillet, "La Escuela Activa: Tesis de grado para optar por el título de Licenciado en Ciencias de la Educación" (unpublished thesis, Bogotá: Escuela Normal Superior, 1937).

42. República de Colombia, *Memoria del Ministro de Educación Nacional* (Bogotá: Prensas del Ministerio, 1951), 5.

43. República de Colombia, *Memoria del Ministro de Educación Nacional* (Bogotá: Prensas del Ministerio, 1949).

44. L. López de Mesa, *Perspectivas culturales* (Bogotá: Universidad Nacional, 1949), 89.

45. See, Herbert M. Kliebard, *The Struggle for the American Curriculum 1893–1958* (New York and London: Routledge and Kegan Paul, 1987).

V

Asia/Asia Minor

11

A History of the Present: Chinese Intellectuals, Confucianism and Pragmatism

Jie Qi

Transvaluation of all values.

—Friedrich Nietzsche

Transvaluation of all Chinese & foreign values.

—Hu Shih

Overview

The chapter investigates the way in which the modern Chinese intellectual has been constructed during and after the May Fourth Movement occurring in 1919 when Dewey started his trip to China. First, it argues that the introduction of Dewey's pragmatism into China was a break in modern Chinese intellectual history. It shifted the notion of intellectuals from a model of Confucian collectivism to one of individualism. Second, this chapter explores how the "New Cultural Movement" brought revolutionary changes in Chinese language, Chinese literature, Chinese culture, and educational systems into a relationship with the reading of Dewey. For example, the main purpose of school education was to prepare children for the governmental servant examination before the New Cultural Movement. However, it shifted to teaching children the "real" tasks that they need in their late life after 1919. Third, the chapter demonstrates that the interpretation and application of Dewey's pragmatism in modern China shifted over time and was interpreted as a traveling library and so brought about reform in thought in a non-confrontational yet critical manner. The shifts of interpretation of Dewey's thoughts caution us to be aware of the "regime of truth." The chapter concludes by pointing out that Dewey's pragmatism assembled into the formation of the particular Chinese modernity. The Chinese modernity is a political issue and it is associated with multiple dimensions.

John Dewey's work was first translated into Chinese in 1912. For near a century, his ideas have been read, interpreted, and applied within changing social, cultural, and political contexts of China. Dewey has been recognized more as a philosopher than an educator in China. Five of the most representative Chinese intellectuals during the twentieth century—Hu Shih, Feng You-lan, Tao Xing-zhi, Jiang Meng-lin, and Guo Bing-wen studied under John Dewey at Columbia University.[1] For the last century, no other Western theorist was as controversial as Dewey in China. Recently, many scholars in China have started to examine how to use Dewey's ideas to solve current Chinese educational problems. However, this study does not try to test how Dewey's theory is useful for current Chinese education phenomena; rather, it seeks to investigate the modernization in China as a cultural process in which Dewey was both hero and antihero in the formation of cultural patterns and modes of life for Chinese to live. My analytic approach to Dewey in this chapter is to place Dewey within *traveling libraries*, that is, I consider how Dewey has become assembled and reassembled with other ideas, institutions, and authority structures related to Chinese life.

The relationship between Dewey and China is one of both a *foreigner* and *indigenous*. Dewey is brought into Chinese discourses of modernization in at least three different times, each expressing a different assemblage of ideas and authority structures. The chapter first investigates the way in which the Chinese intellectual and modernity were constructed during and after the May Fourth Movement (also called the New Cultural Movement), which occurred in 1919 when Dewey started his trip to China. The introduction of Dewey's pragmatism into China was assembled with other authors to tell a story of the beginning of modernity in the history of the Chinese intellectual that shifted from Confucian collectivism to an individualism. Second, this chapter explores how this individualism was constructed with Dewey's thoughts on democracy and science as the new Chinese intellectuals sought revolutionary changes in Chinese language, Chinese literature, Chinese culture, and educational systems. Third, the chapter explores the shifting interpretation of Dewey's pragmatism in modern

China through different sets of traveling libraries as Deweyan intellectuals were labeled "reformers" and "humanitarians" from the 1920s through the 1940s, although later their roles reversed as they were attacked for being "worshipers of America" during Mao's regime (1950s–1980s). In contemporary China, John Dewey's pragmatism once again has been reassembled in the discourses of social, cultural, and educational reforms.

Dewey as the Indigenous–Foreign "Second Confucius"

China was undergoing a rapid social change in the earlier twentieth century. The last imperial regime collapsed in 1911. During this period, Japan and some other European countries attempted to colonize China through military strength. An anti-Japanese and anti-European-power movement spread throughout the whole country. At the same time, some intellectuals urged the learning of Western philosophy and theories, especially those from America. Many Western philosophies and theories were translated into Chinese. Among them, John Dewey's theories of social science and education were prominently acknowledged by Chinese intellectuals. John Dewey's thoughts were introduced into China as early as in the beginning of the twentieth century. Hu Shih, Feng You-lan, Tao Xing-zhi, Jiang Meng-lin, and Guo Bing-wen, well-known modern Chinese philosophers, educators, and social reformers studied with Dewey at the Columbia University in the early twentieth century.

Dewey was the first modern Western theorist to visit and lecture in China. Dewey's twenty-six-month visit started on April 30, 1919. Dewey lectured about philosophy, ethics, and sociology more than education. Accordingly, Dewey has been recognized to be a philosopher more than an educator in China. Dewey lectured around China for more than two years. Dewey taught at Peking University, Peking Coeducational Normal University, and Nanjing Normal University. In addition, he visited fourteen provinces to give speeches. He delivered more than 200 lectures. The topics of these lectures included politics, philosophy, sociology, education, ethics, end so on.[2] Dewey's speeches were translated into Chinese and published in journals and books. Those books were reprinted 10 times in runs of 10,000 copies, and were sold out almost immediately.[3]

There is no doubt that there was a widespread absorption of Dewey's ideas during this period. His popularity and the prominence of his presence made his ideas a widely consumed commodity. The focus of this chapter is of course not on the direct effect of Dewey as a public figure or the direct influence of his words as an expert—though he was given high status in the changes that occurred. The widespread and popular nature of Dewey's thinking is important to note as it implies that familiarity with Deweyan thought and therefore the process of creating a new theory was connected through a Deweyan influence that occurred simultaneously in a multitude of locations and by a variety of scholars.

Dewey's pragmatism was quite different from the previous thoughts such as Confucianism—which was central to the history of Chinese thought. Confucianism, generally speaking, prescribes ethical roles for the society, community, and family. It tells people how to do and act. However, Dewey's thoughts, especially, his ideas of democracy and science enabled the Chinese intellectuals to question and problematize the taken-for-granted Confucian logics and status quos.

For more than 2,000 years, Confucianism has deeply fascinated the Chinese, especially up to the end of the nineteenth century. It advocates that everything can be classified into sets such as the ruling king/ruled masses, the ruling male head of the family/ruled family members, the ruling male/ruled female, the ruling teacher/ruled student, and so on. It would be difficult to find the "self" in the Confucian doctrines. The Confucian sense of the "self" is not "superordinated or individuated, but is rather a complex of roles and functions associated with one's obligations to the various groupings to which one belongs."[4] Here the self only exists in relation to the hierarchy or as part of a greater order. The self-determined individual is not idealized but rather seen as a social aberration. The virtue for the ruled class is to obey the ruling class. It has been believed that the society would be harmonious as long as this order was obeyed. This is what Confucianism means by humanness, human-heartedness, benevolence, and goodness. Up to the May Fourth Movement, the Chinese intellectuals were discursively constructed as practitioners to diffuse such Confucian beliefs with their intellectual activities and the role of educators was to discipline children to cultivate such Confucian virtue and to be human.

The intellectuals assembled Dewey's ideas into the particular Chinese social and historical conditions and produced the Chinese-style pragmatism—social pragmatism, that is, a social science for social reform movements. The popularity of Dewey's pragmatism lay in its adaptability to form a compatible relationship with the Chinese cultural, social, and political situations. Dewey's thoughts in China, once assembled with other Chinese thinkers, do not appear much Western oriented but rather related to an indigenous Chinese way of living. Dewey's ideas practiced in China were as the concept rather than the origin. Unlike the Chinese communism and socialism that advocated social revolution, the Chinese Deweyan social pragmatist reforms aimed at patching up the Chinese social, cultural, and political conditions instead of the destruction of the society and the overthrow of the regime. As I discuss later in the chapter, these social pragmatist reforms started with language since it was believed that the classic Chinese language hindered the Chinese social progress.

Generally speaking, the American progress is associated with the European Enlightenment that places itself as the most civilized among peoples with the science of modernity as its mode of change.[5] However, the notion of the Chinese progress did not embody such European ideas of Enlightenment or civilization but revolution. The process toward modern China involved revolutions in language, literature, education, social system, and so on.

At one level, the introduction of Dewey gave legitimacy to Chinese intellectuals by linking them to modern Western theories considered as "modern." As Hu Shih explained: "Since America is an industrialized, democratic and equal society, the radical social reform movement is not necessary; however, our Chinese society needs to be reformed at all fields. We have to adopt the pragmatism to our social reform."[6]

During this period, the notion of the intellectual shifted and a new generation of modern Chinese intellectuals emerged. Hu Shih, representative of this new type of Chinese intellectual, did not want to participate in government and political affairs before he started to study with Dewey at Columbia University. He believed that in order to be an intellectual he had better do just as Confucius said: "busy oneself in the classics and ignore what is going on beyond one's immediate surroundings." However, Hu Shih argued that he gradually changed his views when studying at Columbia University. Hu Shih assembled Dewey's ideas into his social reform theory: "We believe that the single individual is constructed by multiple social factors. Therefore, in order to reform the society, we have to first revamp those powers which constructed the society and individuals. Thus, social reform is individual reform."[7]

Reformist Chinese intellectuals, especially Hu Shih, modified Dewey's pragmatism into a theory for social reform movement. The movement was called "The May Fourth Movement." The May Fourth Movement was a rupture in Chinese history. It was an anti-feudal, political, and cultural movement that appealed for social, political, and educational reform. For Hu Shih, political activism was a way for intellectuals to exercise power. Many other Chinese intellectuals joined with Hu Shih immediately as they became leaders of the Movement that expressed, meaningfully, Chinese intellectuals taking a stand together in the political stage that sought pragmatic reforms.

The May Fourth Movement broke out in Beijing three days after Dewey's arrival in 1919. Dewey's presence in China, coincidently, urged this movement to spread nationwide. Dewey was called the father of science and democracy during the May Fourth period in China. However, Dewey cautioned that his theories were not universal and therefore some of his ideas may not fit the Chinese particular situation completely.

While Dewey thought of the need to differentiate in the use of his ideas, he also thought of democracy and science as universal that were fundamental for human progress. For Dewey, as Roberto Unger has discussed:

> The idea of a science of human nature or of morals that would lay bare the basic laws of mind or of behavior . . ., the idea of a set of constraints rooted in practical social needs to produce, organize, and to exchange, . . . [and] the idea that these transformative constraints had a certain cumulative direction of their own.[8]

The gist of this discussion is that we have to be skeptical about truth and knowledge. There is nothing essential about human nature, but all is historically and socially constructed. This is at odds with Confucius and so set in motion a process of rethinking a vast number of ideals and social structures that had been created and accepted based on the widespread acceptance of Confucian concepts. The earlier twentieth century was a turning point in China. The anti-Western occupation movement became fierce, but at the same time Chinese intellectuals gave a sacred status to Western philosophy and theories, especially those from America.

Dewey was awarded the "Second Confucius" prize by The University of China in 1920. The word, Confucius used here, does not mean that Dewey's thoughts are similar to those of Confucius but rather that he was a thinker as great as Confucius. The notion of Dewey's democracy and science enabled Chinese intellectuals to rethink the Confucian rationality such as the hierarchical human relation and the harmonious spirit.

There were also shifts in the concept of the individual. The assemblage of Dewey's ideas into Chinese social reform movement went unhindered by his identity as American or foreigner. Certainly his ideas were derived from Western thought, but initially there was no apparent connection made between Dewey and American imperialism. It was only when a new ideology—Maoist Marxism—was introduced—one at odds both with pragmatism and with the Confucian order—that such denouncements were made. This was not led by intellectuals, but rather by the political regime.

The concept of the indigenous foreigner is helpful for explaining this lack of a Cino-centric critique. Dewey is not the "great author" for the Chinese intellectuals. The "globalized" pragmatic ideas could be accepted since they did not form a rigid set of rules or standards that would seek to homogenize the world and replicate American society, but rather provided an understanding that lent itself to reapplication and adaptation. Confucianism could not be so easily introduced to another culture as it is far too specific about roles and social order. By thinking through Dewey's methods, however, the Chinese began to participate in a process that was also occurring elsewhere—acceptance of scientific thought processes that brought a particular modernity into China through thinking about science in organizing daily life. Dewey provided an exemplar or icon in which to render the ideas being formed through similar thought processes around the globe comprehensible to Chinese intellectuals. This paved the way for participation in other Western discourses.

The Shift of the Notion of the Intellectual

Chinese intellectuals were, until the introduction of Deweyan thought, constructed according to Confucianist rationality. For a long time, the final goal for intellectuals was to be trained as official servants to support the regime, as in one of Confucius's well-known proverbs, "studying is for the purpose of being official servant." As I discussed in the previous section, Confucian society is a hierarchical one. What was considered virtue was that one has to be obedient to the superior. Hall and Ames have pointed out clearly:

> The Confucian project is to create community as an extended family. And family relationships are resolutely hierarchical. Ideally, the effects

of hierarchy are meliorated by the processive conception of person. The performance of different roles and relationships enables persons to give and to receive comparable degrees of deference.[9]

It was considered that deference is the way leading to harmony. Harmony was very important in the Confucian society. Confucius said in the Analects:

- Exemplary persons seek harmony not sameness; petty persons, then, are the opposite.
- Exemplary persons associating openly with others are not partisan; petty persons being partisan do not associate openly with others.
- Exemplary persons are self-possessed but not contentious; they gather with others, but do not form cliques.[10]

The words, "harmony" and "self," are not as same as Dewey's. The construction of such Confucian harmony and self was for the sake of the totalistic community. It was believed that as long as those below respected and obeyed those above (i.e., their hierarchical superiors) society would remain in harmony. What the above said was the absolute truth. "Confucianism taught that virtue and power come from a single source."[11] Therefore, questioning authority and truth was considered treason. This type of harmony was not achieved through "mutual adjustment" but through "the appropriate focus."[12] The individual had to have a proper understanding of his or her position and act appropriately in order to achieve such harmony. The individual self was neither distinguished nor appreciated. Everyone belonged to a social group, for emample, a family, a neighborhood, and a village as Hall and Ames have pointed out: "The Chinese understanding of the individual's relationship to authority is distinctly non-adversarial, rooted as it in the metaphor of family."[13] The concept of the self/individual did not arise.

Up to the beginning of the twentieth century, intellectuals in China could be divided into two different types: one type succeeded in becoming official servants and the other did not succeed in official circles. The former worked for the regime and accumulated wealth. The Chinese phrases, "yijin huanxing" (*returning home after making good*), praises the successful Confucian intellectuals. It was understood that the good learner would become an official servant and to be a servant was to make fortune. However, the latter, those who did not become servants tended to detach themselves from reality. They indulged in literature. Although they were dissatisfied with reality, their voices were drowned by the Confucian authoritarianism. The virtue for these intellectuals was considered to be "qinggao," that is, standing aloof from politics and worldly consideration.

Even though the above two types of Chinese intellectuals seem to be dissimilar, neither type was able to escape from the construction of Confucian collectivism. Working for the regime or keeping away from the regime, both types of intellectuals restricted themselves in the hierarchical discourse. Tenzi's (emperor) words were unquestionably considered "tianjing diyi" (*right and proper*). It was impossible for the Chinese intellectuals to think and question about the truth and power. Eisenstadt has pointed: "The non-political elites tended to view themselves as being on a par with and even superior to the political authorities in the political realm, and they very often viewed the political authorities as potentially accountable."[14] Therefore, both types of intellectuals tended to be "mingzhe baoshen" (*be worldly wise and play safe*) rather than criticizing the politics, society, economics, and so on.

At the beginning of his stay in China, Dewey was surprised at the indifference of the Chinese traditional intellectual toward sociopolitical issues. Dewey often presented the following two episodes during his speeches in China in order to encourage Chinese to pay attention to sociopolitical issues.[15] The Chinese intellectuals placed Dewey in the critical position as agents of modernizing and social change. As the conceptual personae, Dewey was able to speak in such a critical way whereas the earlier Chinese intellectuals would not be able to speak due to their indigenousness.

In May 1919 when Dewey was in Shanghai, he asked a Chinese scholar about the Japanese invasion upon Manchuria. The scholar said: "Well, this is Manchurian business but nothing to do with me." The other episode happened in Beijing. One day when Dewey walked back to his dwelling from Qinghua University, he saw a person who was hit by a horse wagon and injured. However, many Chinese pedestrians just passed away. Finally, a foreigner took the injured person to the hospital. These happenings made Dewey ponder about how this Chinese nature had been constructed. As he continued his travels, Dewey mentioned this in a lecture given in the Province of Shangxi that the reason why the Chinese took little interest in sociopolitical and other issues was not because they are cold-hearted or egoistic. Dewey emphasized that the construction of the Chinese nature was associated with mutable factors, for example, Confucianism rationalities, agriculture, lifestyle.[16]

John Dewey's ideas joined with the Chinese intellectuals became a conceptual personae during the May Fourth Movement. The introduction of Dewey's notion of science and democracy were reassembled by Chinese intellectuals to rethink the relationships between politics, society, and themselves. The means of this reassemblage of Dewey's ideas in China were clearly as traveling library through which the May Fourth Movement made intelligible its programs. Dewey did not suggest or participate in the social reform movements as it was instigated by readers and interpreters that brought his ideas into its programs.

New Intellectuals, New Language, and New Literature

During the social movement, Dewey was no longer merely Dewey but was an embodiment of a cultural pattern of ideas. As China's new intellectuals started to take an active part on the sociopolitical stage, they began to question the Chinese language. They sought to convert the Chinese language to a tool that could be able to express modern thinking. During the May Fourth Movement, "modern" is meant to be anti-Confucianism and present oriented (e.g., facing the reality of the present society and participating the social reform movements as a means of undoing particular Chinese traditions). The term, present-oriented society, was used to contrast with the past-oriented society, which was conservative and retrospective. Until the language and literature reform movements of the May Fourth Movement, the Chinese literature was full of reminiscences and sentimentalities. It avoided talking about the everyday life and social and cultural issues of the present.

The major linguistic reform started from the early 1910s and the May Fourth Movement was the climax. Up to the May Fourth Movement, the Chinese language consisted of two forms: *wen yan* and *bai hua*. The former, *wen yan*, is classical language and the latter, *bai hua*, is vernacular language. Written Chinese was completely different form spoken Chinese in grammatical structure and vocabulary. Written Chinese, *wen yan wen*, was written in classical style. *Wen yan wen*, the "esoteric classical writing [was] intelligible only to scholars."[17] The traditional scholars not only had to be able to read but also to write in *wen yan wen*. It took a long time for scholars to become familiar with *wen yan wen*.

The new intellectuals considered *wen yan wen* to be an obstruction that restricted intellectuals' thinking abilities due to its complicated sentence structure, ancient vocabulary, difficult orthography. Hu Shih called *wen yan wen* a "half-dead" and "nonscientific" language:

> A dead language can never produce a living literature; if a living literature is to be produced, there must be a living tool. . . . We must first of all elevate this [vernacular] tool and make it the acknowledged tool of Chinese literature that totally replaces that half-dead or fully dead old tool. Only with a new tool can we talk about such other aspects as new ideas and new spirit.[18]

Thus, the urgent matter was to find a way to reform such a half-dead language. New Chinese intellectuals considered Dewey's thoughts of science and democracy as of inseparable value. They believed that reform needed a scientific and problem-oriented approach. This approach is not only a means of creating and testing new methods, but also a means of critiquing the previously accepted with an appeal far more effective than simply advocating an alternative. Schwartz has summarized the "scientific" method for classic language reform:

> The methods of "science" could . . . be used "to undermine the credibility of orthodox histories and the historical foundations of scripture." One of the most effective ways of removing the dead hand of tradition was to dissolve the factual claims of the myths which supported that tradition. In the end this critical liberation of historical studies from the burden of certain fundamentalistic and conventional ways of viewing the past was to be taken up by many other scholars of "national studies" . . .[19]

Hu Shih's advance of language reform assembled with Dewey's idea of science. Hu Shih believed that new and common language was a technology of communicating new thoughts and that language could lead to educational reform. In 1917 Hu Shih posed the following eight detailed proposals for language reform:

(a) Talking about the real world or substance.
(b) Not imitating classical sayings.
(c) Paying attention to grammar.
(d) Not adopting a sentimental pose.
(e) Discarding hackneyed phrases.
(f) Not using the classical allusions.
(g) Not using antitheses.
(h) Using common words and sayings.[20]

This proposal attempted to reform the unpopular and difficult classic Chinese language to a simplified popular language. The classic Chinese language was completely different from spoken language. It was detached from the real life and society. It was the symbolic power. The reform of the classic language unified the written language and spoken language into one form and it therefore enabled the masses to become literate easily. The proposals for language reform along with educational reforms, which I will discuss later in the chapter, can be viewed either in a Western or Chinese particular perspective of "modernity." On the one hand, these reforms articulated the idea of freedom and democracy that assembled some of Dewey's humanist ideas but at the same time these proposals provided the Chinese particular "present-oriented" approach to modernity. During the late 1910s and the 1920s, "modernity" connotes the meanings of being opposed to the past orientations of classic Confucianism, taking interest in the current Chinese social issues and forward-looking search for the future of China. Language reform, therefore, was the first step toward to the "modernity" as the classic language was considered the primary hindrance to the spread of education and social progress.

Hu Shih marked a milestone in the history of the Chinese language reform. He worked with many other leading China's intellectuals to give leadership to this language reform, such as Qian Hai-tong, Fu Si-nian, Liu Ban-nian, Zhou Zuo-ren, Zhu Wo-nong, Yu Ping-bo, Zhu Xi-zhu. These scholars took the lead in writing in the style of *bai hua*, the vernacular language. Such writings have been called *bai hua wen*. Simplified Chinese characters, a new way of writing, a new structure of grammar and phonetic notation were applied to their writings. Even the writing style changed from vertical lines to horizontal lines since it was believed that the horizontal lines were more modern, scientific, and convenient—for one thing, the right-handed scholar can now rest his wrist on the page without smudging what has already been written.[21] Many novels, operas, dramas, poems, and essays written in vernacular language sprang up. Simultaneously, the number of periodicals increased dramatically. There were nearly 200 new periodicals founded around the time of June 1919.[22] These new writings introduced and created new words that did not exist in the classic language, for example, "liberty," "individualism," "democracy," "revolution," "citizen," "country," "republic," and so on. These words brought new thoughts and values to Chinese people. It enabled the emergence of the new China's intellectuals, the new China's youth, and the new writers.

It was the very first time that individualism and subjectivism appeared in Chinese writings. Since the classic literatures were full of Confucian moralizing discourse, the writer's personality, life experience, and feeling were excluded from the classic literatures. The purpose of writing for the classic writers, generally speaking, was to recite the Confucian ethics and morality and these classics were writing for Confucian scholars to read. However, the purpose of writing for the new writers was to enlighten people to rescue the Chinese society. Leo Qu-Fan Lee, a professor of Chinese Literature, has pointed out clearly through the analyses of the work of Lu Hsun, one of the most famous new writers:

> Lu Hsun himself acknowledges at different times that two significant impulses lay behind his fictional writing. He declared that its purpose was to enlighten his people and to reform society. . . . But he also admitted that his stories were products of personal memory: he wrote because he had been unable to erase from memory certain aspects of his past which continued to haunt him. Thus, in his fictional output he endeavored to combine artistically a private act of remembrance of things past with a public concern for intellectual enlightenment. He attempted to rearrange recollections of China's national experience, thereby making it less ego-centered, as in most early May Fourth literature, and more significant to his readers.[23]

The purpose for the new writers writing new novels is to reform custom, politics, religion, morality, the Chinese characteristics, and the Chinese society.

The new Chinese intellectuals' struggle for language reform was crowned with great success. On January 1921, the Ministry of Education abolished the use of *wen yan wen* at schools. Instead of *wen yan wen*, the vernacular language, *bai hua*, became a national language in both written and spoken form. The significance of this language reform movement was that it fostered the emergence of the national language in the 4,000-year Chinese history. It enabled the Chinese to think, talk, and write in the same language that provided a new sense of belonging and way of thinking of the collective identity of the nation. It also led to educational reform in the 1920s.

In sum, Dewey's thoughts on science and democracy reassembled in the discourses of language reform. The diffusion of vernacular language produced new literature. Moreover, the new literature introduced new knowledge, thoughts, custom, value, and so on. All these factors, together, provided Chinese people with a new reasoning: a new way of understanding. A new type of the citizen emerged.

The Memento of Pragmatism in China: Educational Reform after the May Fourth Movement

In the 1920s, Dewey's pragmatism reassembled in the discourse of school reform. Especially, "child-centered education" became the principle that enhanced educational reform after the May Fourth Movement. The new intellectual leaders, Hu Shih, Tao Xing-zhi, Jiang Meng-lin, and Guo Bing-wen who studied with Dewey at Columbia University, initially were educators. The experiences of study with Dewey provided them with a chance to think of themselves as modern Chinese thinkers, social reformers, politicians, and so on. However, they still did not forget the field of education.

The call for educational reform included three aspects: educational philosophy, educational system, and pedagogical issues. Under the Confucian system, education is for the sake of preparing children for the governmental servant examination. It is called *ke ju zhi*, which means imperial examination system. This was a system for examination, academic degree, and governmental service. If students did well in the examination, they would get positions in the government. The higher the score, the better the position acquired. Therefore, learning was the gate to pass through into official circles. The Confucian belief regarding the relationship between the teacher and the student is that the teacher is the sage and the student has to obey the teacher. The child was required to be well mannered, sitting straight on the stool, reciting the classics. The child should not show any emotional feelings. No laughs or mutual communication were permitted in the classroom. Cramming students with knowledge, mechanical memorizing, and examinations, above all, were the modus operandi at Confucian schools, which were criticized by the new intellectuals. The child is no flesh, no blood, and "half dead." Lu Hsun, one of the new China's leading intellectuals, said in his famous essay, "A Madman's Diary" that it is urgent to rescue the child from the Confucian educational system. At the same time, Tao Xing-zhi, one of Dewey's students, appealed to free the child from the deadly Confucian educational system:

(a) Free the child's eyes.
(b) Free the child's brain.

(c) Free the child's hands.
(d) Free the child's mouth.
(e) Free the child's space.
(f) Free the child's time.

He believed that in order to draw the child's creativity and individuality it is necessary to give the child autonomy.[24] This educational reform assembled some of Dewey's ideas of child-centered education. However, it is different from the Dewey's approach. Dewey's concern is the issue of curriculum, whereas the Chinese reform does not associate much with curriculum but mainly the physical surroundings. For example, the chairs for students were switched from stools to chairs. This was the first time that the Chinese intellectuals spoke for the child.

Educators tried to find a scientific way to free the Confucian child. They came to a scientific approach to reform language education as it was believed that the reciting of the classics was most harmful to the child and it was the stumbling block to the progress of the society. Many experiments in writing and reading Chinese characters were carried out during the later 1910s whereas experiments on language teaching became popular after the May Fourth Movement. Ai Wei has summarized the major results of these experiments:

(a) Simplified Chinese characters are easier to write and memorize.
(b) Chinese characters designated for daily use were selected for seventh grader to twelfth graders.
(c) Writing or reading horizontally was faster than writing or reading vertically.
(d) Children read interesting stories much faster than reading classics.
(e) Reading silently helped understand the contents more than reading aloud.
(f) Children could write better on free-writing than given-topic-writing.[25]

These experiments led to later educational reforms. Simultaneously, Dewey's ideas on democracy and education fascinated many intellectuals. Tao Xing-zhi explained: "Democracy and education is a very important approach to reforming the Chinese society. . . . It enables different people to enjoy democracy. . . . I will create a democratic society with democracy and education."[26] Tao Xing-zhi was the first one to start applying Dewey's style of experimental school in China.

The intellectuals' efforts toward educational reform effected official changes in educational systems. In 1921, the Ministry of Education decided to abandon *wen yan wen*, the classic language, at schools. As I mentioned in the previous section, this reform created the national language and it enabled the Chinese to think, talk, and write in the same language. It also freed children from reciting classics.

In 1922, the Ministry of Education further laid down a regulation to introduce the American 6-3-3 system into schools.[27] The Ministry of Education prescribed the following seven guidelines for educational reform:

(a) Suiting the needs of the social evolution.
(b) Bringing the ideas of the democratic education into effect.
(c) Seeking to develop individuality.
(d) Becoming concerned with the economy.
(e) Becoming concerned with home economics education.
(f) Making education universal.
(g) Paying special attention to local needs (my translation).[28]

These regulations clearly embody ideas that relate to Dewey, such as child-centered education, learning from real life, interest in relation to training of the will, and so on. The link to Dewey becomes indirect here. It is not the words or suggestions of Dewey, but the general climate of values (i.e., changeability, analysis, individuality) for the creation of such policies. The concepts of changeability, analysis, and individuality in the discourse of the modernity constructed the Chinese "modern" citizen. The idea of "changeability" does not exist in the Confucian doctrine. Confucianism regards "truth" and "knowledge" as universal and immutable. However, the modern citizen is skeptical about "truth" and "knowledge." Analysis is another important aspect that forms the modern citizen, that is, the modern citizen has the abilities to problematize the status quo and to solve problems. The Chinese individuality, though, does not place much emphasis on individual freedom, but stresses the importance of the individual rights within a group, community, and nation.

In the Chinese discourse, the word "individual," in the context of "individual rights within a group" means "individual person." Whereas it refers to "group" when individual is used in "individual rights within a community." Moreover, it also means "community" when it appears in "individual rights within a nation." In Chinese, "group" means small nonofficial or officially registered organizations, for example, family, class, neighbor. On the other hand, the word "community" in Chinese does not have the English meaning of "the Chinese community" or "the Japanese community" that refers to a normative sense of collective belonging. It means officially registered organizations, for example, factory, school, residential district, village, city, prefecture.

In 1927, the minister of education issued a command that required all public university primary schools in the Province of Jiangsu to be changed to experimental schools. Thus, experimental schools became popular throughout the whole country. The purpose of this reform was to use a scientific approach to solve some educational problems. For example, scientific language learning and teaching experiments were conducted in these experiment schools. The experimental school movement became more and more popular until the end of 1940s.

The series of educational movements were to change the Confucian child to the Chinese version of the modern child: they are allowed to ask the teacher questions; they can write out their own chosen topics; they do not have to always recite the reading; they can play during the recesses (there was no recess time before the school reform as it was believed that the child had to study all the time when they were at school); they can enjoy reading children's stories at school. The child is allowed to act like a child not a "little-adult."

Dewey's thoughts on democracy and education were defined as the scientific approach to educational reform in China from the May Fourth Movement onward. However, the purposes of education for Dewey and for the China's new intellectuals were different. Lawrence Cremin has noted:

> . . . education began as part of a vast humanitarian effort to apply the promise of American life—the ideal of government by, of, and for the people—to the puzzling new urban-industrial civilization that came into being during the latter half of the nineteenth century.[29]

Dewey's project of education is to suit education to life; however, education for the new Chinese intellectuals was to reform the Chinese society and to remold the Chinese life. Furthermore, Dewey's idea of the relation between the school and real life was not adopted by the Chinese educators in the following two aspects. First, Dewey tried to make schools like real-life conditions and let children learn the real-life skills in the classroom, that is, the imitated environment. The Chinese experiment school students learned outside school, for example, in the real environment. The Chinese educators believed that the real-life environment is the best school. Tao Xing-zhi, the famous Deweyan educator, advocated that the living environment is the best education, the society is the best school, and teaching and learning help each other.[30] The Chinese educational reformers believed that it was important for students to follow such a learning process—practice, cognition, and practice. Second, Dewey thought of people who were living in the urban environment and therefore Dewey's experiment school was established in the city. On the other hand, the distinguishing characteristics of the Chinese experiment schools were placing emphases on agricultural education. The Chinese experiment schools were founded in the countryside as the Deweyan educators considered China to be an agricultural country and most of the countryside people were illiterate. One purpose of the experiment schools in China was to spread education in the rural districts. Thus, it was believed that the rural districts were the best real-life environment for education. Teacher and students built the experiment school buildings by themselves. They also cultivated the fields and did their own cooking. Students taught the village people reading, writing, and counting. Students learned from whatever was around them. Here again local discourses set in motion by Dewey were shaped by China's specific circumstances.

The Fickle Destiny of the Indigenous Foreigner

Dewey was the conceptual persona for the modern Chinese intellectuals. However, perhaps no country has treated Dewey in such a fickle way as China has. Dewey's work was first translated into Chinese in the year of 1912. Hereafter, in China, he was admired as the father of "democracy" and "science" as well as a great philosopher, humanitarian, and educator. Although his ideals were completely different from those of Confucius, he had one thing in common. Confucius and Dewey were both criticized and denounced during Mao's regime (1950s–1980s). Up to the end of the 1940s, Dewey's thoughts assembled with other Chinese scholars in the changes associated with social, language, literature, and educational reform movements. Nevertheless, his destiny took a tragic turn with the foundation of the People's Republic of China in 1949.

Since the year 1949, the new China started the nationalization of properties. Land, factories, banks, and all other such entities became public owned or state owned. Private ownership of production was completely abandoned. All the people in the country were working for the government or the public. Intellectuals were no exception.

The nationalization of the intellectuals became more and more severe. The nationalized intellectuals were expected to tell the "regime's truth," that is, the "knowledge" and "truth" produced by the government. Questioning such "knowledge" and "truth" was prohibited. Critical thinking abilities were considered threats to the regime. The theories, which enable people to think and question about the status quo, to reform the society, were inevitably discounted or discarded.

Dewey's theories were the first target of the criticism after the foundation of the new China. In May 1955, *China Education*, a nationwide monthly journal, published an editorial article to criticize Dewey's pragmatism. Hereafter, Dewey became the symbol of capitalism and American imperialism.[31] Hu Shih was also criticized for being a servile follower of the American imperialism. The critics blamed him for the following mistakes:

(a) Reactionary philosophy.
(b) Against Marxism.
(c) Incorrect historical view.
(d) Incorrect thoughts toward literature.
(e) Historical idealism.
(f) Incorrect historical view of literature.
(g) Wrong textural criticism.
(h) Destroying education.[32]

Dewey's notion of science and democracy were labeled as "pseudo-science" and "pseudo-democracy." Up to the end of the Mao's regime, Dewey's thoughts were under the attack of criticism for over twenty years.

However, "it grows again when the spring breeze blows," as the Chinese proverb goes. In the middle of the 1980s, Dewey's thoughts were once again to be studied by the "second generation of new China's intellectuals." Since the early 1990s, the Ministry of Education of China has emphasized the importance of the study of Dewey's thoughts on its guidelines for national education. Dewey continues as an indigenous foreigner during the turbulent twenty-first century with China and China's intellectuals.

In Closing

My intention in undertaking this study was not to take John Dewey's pragmatism as a simple substitute for Confucian discourse. I sought to

examine how Dewey's ideas have been read, interpreted, and practiced under different social and historical conditions and how multiple factors engendered the new concept of the Chinese intellectual.
I first looked at the construction of Confucian intellectuals and how
Dewey's thoughts were connected with Confucian intellectuals to
rethink their intellectual activities with relevance to their sociopolitical
world. The notion of the intellectual shifted dramatically during the
period of the late 1910s and early 1920s. Being an intellectual no
longer meant an official scholar or a person who is free from the vul-
garity of political involvement, but means the ability to criticize the
sociopolitical world and to take a leading role in various reform
movements.

I next investigated how the new movement among intellectuals
strove to reform language and literature. Language and literature are
not a mixture of words, phrases, and grammar but the effects of all
social, political, and historical discourses. The reformed language
and literature provided the Chinese with a new possibility for reasoning and communication of ideas. The new intellectuals used new language and created a new literature to tell a new story that was
different from the Confucian story. The new story told a new "truth"
and "knowledge" and such new "truth" and "knowledge" discursively
constructed the notion of the particular Chinese modernity. Note here,
that this new knowledge was not created by Dewey, but by means of
the—modified, indigenized—tools he provided.

The new language and literature further affected a series of
educational reforms. This reform included three aspects: educational
philosophy, educational system, and teaching methodologies. The
significance of these reform movements was that it changed the
Confucian child into a modern individual. However, these educational reform movements, which were carried out after the May
Fourth Movement did not follow the principle of Dewey's pragmatism
exactly. They were the result of multiple effects, for example, Dewey's
thoughts, the China's reformers' own experiences, the particular Chinese
social, economic, and political situation.

The particular Chinese cultural, social, and political circumstances
constructed the Chinese notion of individuality, community, progress,
and change. The Chinese individuality focuses on the individual right
rather than autonomy. In the Chinese discourse, community is an officially registered totalistic organization and therefore it is much different from the community in the Western discourse. The notion of the
Chinese progress is not entailed in European ideas of Enlightenment

and civilization but in cultural revolution that produces change and discontinuity in the Chinese context.

Finally, I traced the contingency of Dewey's ideas that have been read and interpreted in modern China. During different historical periods, Dewey's ideas have been interpreted completely differently. The hero is not always the hero. It is socially constructed and historically contingent. The rise and fall of Dewey in modern Chinese history tells us a story of the "regime of truth." We have to be skeptical about how the truth and knowledge have become self-evident.

Dewey's pragmatism assembled into the formation of the particular Chinese modernity. The Chinese modernity is a political issue and it is associated with multiple dimensions. On the one hand, when modernity is articulated as a set of progressivism and conservatism, modernity is understood as the ideas of evolutionism and progress, the negation of the absolute power, the respects for science, technology, and individuality. However, when modernity is considered a set of Eastern and Western, the Western modernity is seen as the adoption of the Western civilization, governing systems, lifestyles, and so on. The Eastern modernity is "combined with very strong anti-Western and anti-Enlightenment ideologies . . . and attempt to appropriate modernity on their own terms; and the total reconstruction of personality and of individual and collective identities by conscious human, above all political action, and the construction of new personal and collective identities of entailing the total submergence of the individual in the totalistic community."[33] This type of the Eastern modernity ran amuck during Mao's regime.

Notes

1. The order of Chinese names, last name and first name, is used throughout the chapter.

2. Qing Yuan, *Duwei yu zhongguo (Dewey and China)* (Bejing, China: Renmin Chubanshe, 2001).

3. Zhong-hui Dan, *Xiandai jiaoyu de tansuo—Duwei yu shiyong zhuyi jiaoyu sixiang* (Contemporary Education: Dewey and Pragmatism) (Beijing, China: Renmin Jiaoyu Chubanshe, 2002); Bao-gui Zhang, *Duwei yu zhongguo (Dewey and China)* (Shijiazhuang, China: Hebei Renmin Chubanshe, 2001).

4. David H. Hall and Roger T. Ames, *The Democracy of the Dead: Dewey, Confucius, and the Hope for Democracy in China* (Chicago: Open Court Publishing Company, 1999), 209.

5. S. N. Eisenstadt, *Comparative Civilizations and Multiple Modernities*, Vol. II (Leiden: Brill, 2003), 493–518.

6. Shih Hu, *Hu Shi Jiangyan* (Hu Shih's Lectures) (Beijing, China: Zhongguo Guangbo Dianshi Chubanshe, 1992), 416.

7. Shih Hu, *Shiyong rensheng: Hu Shi suixianglu* (Pragmatism and Life: The Essays of Hu Shhi) (Guangzhou, China: Huacheng, 1991), 291.

8. Roberto M. Unger, *Social Theory: Its Situation and Its Task*, Vol. 1 of *Politics: A Work in Constructive Social Theory* (New York: Cambridge University Press, 1987), 85.

9. Hall and Ames, *The Democracy of the Dead*, 160.

10. Quoted by Hall and Ames, *The Democracy of the Dead*, 193.

11. Charlotte Furth, "Intellectual changes: From the reform movement to the May Fourth Movement, 1895–1920," in *An Intellectual History of Modern China*, ed. M. Goldman and L. O. Lee (Cambridge: Cambridge University Press, 2002), 37.

12. Hall and Ames, *The Democracy of the Dead*, 194.

13. Hall and Ames, *The Democracy of the Dead*, 195

14. Eisenstadt, *Comparative Civilizations*, 291.

15. Bao-gui Zhang, *Duwei yu zhongguo*.

16. En-rong Song, *An Yan-chu quanji* (The Collected Works of An Yan-chu) (Changsha, China: Hunan Jiaoyu Chubanshe, 1992), 278.

17. Benjamin I. Schwartz, "Themes in intellectual history: May Fourth and After," in *An Intellectual History of Modern China*, 97.

18. Quoted by Leo Ou-Fan Lee, "Literary trends: The quest for modernity, 1895–1927," in *An Intellectual History of Modern China*, 158.

19. Schwartz, "Themes in intellectual history: May Fourth and after," in *An Intellectual History of Modern China*, 114.

20. Shih Hu, *Hu Shi koushu zizhuan* (Dictated Autobiography of Hu Shih) (Shanghai, China: Huadong Shifan Daxue Chubanshe, 1993), 149.

21. There were some experiments in writing and reading carried out by Deweyan educators during the late 1910s. The results showed that writing and reading horizontally was faster than writing and reading vertically.

22. Zhong-hui Dan, *Xiandai jiaoyu de tansuo*.

23. Leo Ou-Fan Lee, "Literary trends: The quest for modernity, 1895–1927," in *An Intellectual History of Modern China*, 175.

24. Zhong-hui Dan, *Xiandai jiaoyu de tansuo*.

25. Hong-shun Cao, *Yuwen jiaoxuefa mantan* (Methodologies of Language Education) (Shenyang, China: Liaoning Daxue Chubanshe, 1992); Huang-chu Gu, *Yuwen jiaoyu lungao* (On Language Education) (Beijing, China: Renmin Jiaoyu Chubanshe, 1995).

26. Xing-zhi Tao, *Tao Xing-zhi quanji* (The Collected Works of Tao Xin-zhi) (Changsha, China: Hunan Jiaoyu Chubanshe, 1986), 55–56.

27. The 6-3-3 system was abolished during the period of the Great Cultural Revolution (the late 1960s–the late 1970s) and revived from the early 1980s.

28. Zu-yi Du, *Duwei lun jiaoyu yu minzhu zhuyi* (Dewey's Thoughts on Education and Democracy), trans. H. Cheng and G. Hong (Beijing, China: Renmin Jiaoyu Chubanshe, 2003).

29. Lawrence Cremin, *The Transformation of the School: The Progressivism in American Education 1876–1957* (New York: Vintage Books, 1961), viii.

30. Xing-zhi Tao, *Tao Xing-zhi quanji dierjuan* (The Collected Works of Tao Xin-zhi, Vol. II) (Chengdu, China: Sichuan Jiaoyu Chubanshe, 1991).

31. Shu Chen, "Duwei de fandong sixiang zai xinlixue shang suo biaoxiang de liangge lizi (Two examples of Dewey's reactionary thoughts on psychology)," *Gushi Kao* (Ancient Historical Review), no. 2 (1955): 385–388; Yi-han Gao, "Shiyong zhuyi de fandong sixiang benzhi (The reactionary thoughts of Pragmatism)," *Gushi Kao* (Ancient Historical Review), no. 2 (1955): 541–544; Qu-yuan Hu, "Shiyong zhuyi pipan (Critique of Pragmatism)," *Gushi Kao* (Ancient Historical Review), no. 2 (1955): 287–294; Yue-ling Jin, "Pipan shiyong zhuyi de Duwei de shijieguan (Critique of Dewey's pragmatism)," *Gushi Kao* (Ancient Historical Review), no. 3 (1955): 113–136; Pei Zhang, " 'Xuezhe'—zhengzhi yinmoujia—Hu Shi zai sixiang shang he zhengzhi shang de fandong benzhi ('intellectual'—political intrguer—Hu Shih's reactionary philosophical and political nature)," *Gushi Kao* (Ancient Historical Review), no. 1 (1955): 391–400.

32. Peng He, "Hu Shi zai duidai woguo wenhua chuantong zhong de diguo zhuyi nucai mianmu (Hu Shih—servile follower of the imperialism: Destroying our cultural tradition)," *Gushi Kao* (Ancient Historical Review), no. 2 (1955): 313–324; Te Ma, "Shiyong zhuyi—zui chenfu, zui fandong de zhuguan weixinlun (Pragmatism—the most commonplace and reactionary subjective idealism)," *Gushi Kao* (Ancient Historical Review), no. 2 (1955): 9–28; Guo-en You, "Pipang Hu Shih de zichan jieji weixinrun xueshu guandian he ta de sixiang fangshi (Critique of Hu Shih's bourgeois idealist conception and his way of thinking)," Gushi Kao (Ancient Historical Review), no. 1 (1955): 207–212.

33. Eisenstadt, *Comparative Civilizations and Multiple Modernities*, Vol. II (Leiden: Brill, 2003), 512.

12

Dewey and the Ambivalent Modern Japan

Kentaro Ohkura

Introduction

This chapter illuminates the ambiguity of the Japanese identity by reexamining Japanese studies of John Dewey conducted during the watershed period before and after World War II. Dewey viewed as an *indigenous foreigner* provides an example of the ambivalence of the modern Japan in its struggle in coming to terms with its indigenous and exogenous values over political, social, and educational principles. Thus, this essay is not a case study for introducing the cultural uniqueness of the Japanese understanding of Dewey's ideas, but rather to indicate the ambiguity of the modern Japanese in their understanding of Dewey. This ambiguity played an important role in providing a rationale for the modernization and building of a democratic Japan. Prewar and postwar Japanese scholars constructed their image of Dewey as a kind of *traveling library* that served the modern conditions of their country.

Prewar and postwar Japan were both concerned with consolidation of the Japanese people. It is said that Imperial Japan valued an idealism that supported and encouraged a national spirit and a type of self-identity entwined with the state. But, both pre- and postwar Japan embraced scientific knowledge of the West as a way for talking about, and even "seeing" objects that encouraged a kind of social integration. Thus, both periods can be described as being engaged in the same modern project, that of *nation building*. It is this continuity of history that suggests reconsidering the explanations for how people become consolidated into a nationally "homogeneous" whole at different points in time. The nation-building efforts of Japan emerged from this ambivalence of indigenous/exogenous values over modernization. From the view of history as it persistently continues, a democratic society (i.e., postwar Japan) cannot be simply separated from the totalitarian state of prewar Japan.

This historicity is discussed in terms of the "defeat" in the War and how that impacted what people saw as "real." Postwar Japanese frequently described their experience of this time as a kind of repentance for causing the War.[1] This repentance produced counter narratives that essentially constructed a perceived condition of postwar Japan merely in terms of its opposition to the vales of prewar Japan. This repentant attitude was used in postwar Japan to clearly differentiate themselves from prewar Japanese. This constructed separation from the past helped to produce differences in the readings of Dewey and the views of self-realization and nation building from one period to another.

Finally, the chapter illustrates Japan's continued emphasis on becoming modern a goal that has persisted since the prewar era, but with a particular ambivalence that gives it a distinctive character. Indeed this focus on modernization has played an important role in structuring the public discussions of Dewey in very different ways at different times in Japan's history.

Historical Overview of Modern Japan and Dewey

W. G. Beasley, the British historian, noted that "Japan had no tradition of political theory in the European sense."[2] As he suggested, the dawn of a new Japan (synonymous with *Meiji* Japan) began in 1868 with the "unique" way in which the warriors of Japan restored the emperor to power. The emperor was believed to represent the country's legacy through the lineal succession of their imperial ancestors having only been temporarily removed from power. Unlike the United States, modern Japan could be characterized as being born "ancient." That antiquity did not, however, imply a return to the ancient form of society. Instead, it reorganized elements of tradition with modern knowledge and modern institutions (*wakon yosai*). As a matter of fact, Westerners reminded the Japanese leaders of *Meiji* Japan to not merely view the modernization of Japan as a kind of "Americanization" or through the adoption of Western systems en masse, but rather by reorganizing antiquity and reviewing their cultural traditions in conjunction with the process of modernization.[3]

Although antiquity and modernity represent two opposing concepts, both played similar roles in establishing acceptable conduct for the people of Japan. For example, the concept of time was perceived in both "old" and "new" ways because groups of people in the feudal and Meiji Japan practiced time in two different time frames. Time was regional and indefinite in the feudal era, and national and definite in the modern Meiji period. In the feudal school, an indefinite time frame was employed. Each pupil went to school when one wanted to, and studied at one's own speed under the close supervision of a teacher. In the publicly funded schools of Meiji Japan, by comparison, *all* students were instructed to arrive at school "ten minutes before school begins."[4] This was not merely to reinforce the "standardization" of schooling, but also emphasized a common understanding of time as a form of discipline. The transition from old to new cannot be simply defined as a process of Westernization or industrialization emphasizing "enlightenment," "rationality," and "accuracy." Modernization also involved more subtle forces, such as the manipulation of objects and people through the concept of time, thereby promoting the development of togetherness among the people, and thus encouraging a sense of nationhood.

Along with the emperor's *Oath in Five Articles*, Imperial Japan continued to seek modern forms of knowledge to strengthen the foundations of imperial rule, and eventually inviting Dewey to Japan in 1919. The attitude taken toward Dewey was nonetheless varied among modern Japanese. Dewey's work was not familiar to the academic philosophers of the Imperial University, but was more popular among instructors of private institutions and schools. This contrast between the Imperial University and private schools represents multiple understandings of Japanese modernization and Dewey.[5] The former perceived the adoption of Dewey's philosophy as an importation of Western values, or in short, "Americanization" while the latter characterized it as a form of secularization or democratization.

The rise of the new education movement (which is often likened to progressivism in the United States and the New Education Fellowship in Europe) saw increasing numbers of translations of Dewey's books, essays, and research articles and the New Education movement continued until 1929. It then completely waned as the National Mobilization Law of 1938 gave priority to military requirements for capital, industrial, and trade control. Toward the end of World War II, there remained a view held by some Japanese that learning from Dewey was merely a strategy to make Japan more capitalistic or Americanized.

The postwar period between 1945 and 1969 was the most prosperous time for studying Dewey in Japan than any other historical period. Major reasons for this were the Allied occupation (1945–1952) and the power shift to democracy that pressed for intellectual exchanges between the two countries. In 1952, the John Dewey Society of Japan was established for the purpose of better understanding Dewey as well as American philosophy. It is generally recognized that Japanese educators, who understood Dewey in terms of experimentalism, a learning style based on individual practice, popularized Dewey's approach during this same time period.[6]

As a result, postwar schools were encouraged to address democratization in school and to focus more on the individual student than was the case in prewar schools. This child-centered approach was in large part based on the notion that students gained knowledge through self-reflection and consideration of their individual circumstances in life. With this focus on the student and his/her own life, an experimental, problem-solving approach became the chief tool for structuring learning. Educators were told that learning materials should not be designed with ready-made solutions, but rather should encourage each student to see things in a problematic way "through living and in relation to living," and thereby solve problems in one's own manner.[7]

This experimental approach was also used as a tool for responding to postwar Japan's interest in promoting Japan as a democratic society by encouraging a common understanding that could be reflected through individual practice and natural inquiry. Nonetheless, these scientific tools were employed primarily to reinforce the general values and universal rules that governed the population, rather than the varied individual solutions that Japanese followers of Dewey had intended. Thus, the concept of science in postwar Japan suggested only naturalistic procedures for teleological purposes with little consideration of the way in which it was applied. Postwar Japanese perceived the practice and process of individual learning separately.

A few Japanese scholars of Dewey viewed science in a slightly different way. The problem-solving approach they perceived to be advocated by Dewey enabled them to argue that *all* pupils were capable of learning independently without reliance on some set of general rules. In this view of science, every individual was believed to possess reasoning as a "scientific" tool (i.e., the five senses) that could lead individuals to correct, yet individualized answers. As this scientific attitude became better known among educators of postwar Japan it represented a perspective suggesting that while answers found via the problem-solving approach could be varied, they were based on a reasoning faculty shared by *all*. As Japanese scholars of Dewey realized, the problem-solving approach to learning was therefore democratic insofar as individuals were expected to find their own answers. But, it was at the same time compatible with a meritocracy in a democratic society (where every individual found the shared answer in a competitive environment). This educational perspective and influence thus offered another valuable tool for nationalization.

Deweyan Approach to the Self in Science Education

As suggested in the previous section, science education functioned as a system of knowledge that served to mobilize and integrate the people as a nation. It was not just a methodology for examining objects for practical purposes, but also provided a vision of a modern society where everyone could participate as a productive citizen.

Thus, science education proceeded in prewar Japan under the assumption that every member of society was equally endowed with a natural faculty for understanding others and the universe. Japanese scholars of Dewey were however ambivalent in discussing scientific education. They referred to scientific knowing as exercised by two different approaches or processes more often than as completed by the individualistic use of the natural faculty. This was partially reflected by the way in which the Japanese viewed themselves as being connected to other members of society.

Science education became available to all students in primary school in 1908. In the beginning the Ministry of Education encouraged teachers to employ observation and experimentation, thereby fostering an individual style of learning that was opposed to the more collective, traditional approach to Japanese education (i.e., whereby the student was expected to learn the same content and to deductively "achieve truth"). However, the Ministry standardized the national science textbook as a collection of objective facts in 1911. As soon as it was published, science education came to be seen as an "encyclopedic" type of learning that expected all pupils to share the same view and understanding of the scientific body of knowledge. For example, looking at a page on "sprouts" in the sixth-grade textbook, one finds a common view expressed as a scientific truth that "When it gets warm in the spring, the roots of trees takes nourishment from the soil that holds water. . . ."[8] The normal and primary schools later criticized this approach to truth, and accordingly, condemned selected sections of the textbooks (such as the example above) by arguing that such knowledge was too bureaucratically centralized and too conceptually oriented. As a result, the argument continued, the information presented in these texts would not be familiar to some people in particular regions.[9] In other words, schools during the prewar period disagreed about the specific sections of the science text, and whether they were relevant to all, or should be restricted to certain groups of people. A Japanese scholar, upon returning to Japan after touring Germany and the United States, claimed that Japanese classrooms should adapt the experimental approach to science.[10] The Ministry responded to these calls by attempting to reedit the national textbook, but it was never successfully revised and as a result many primary schools never fully introduced the experimental method.

While primary education extended to nearly a 100 percent of Japanese students by the time of World War I, the school curriculum was not consistent for all pupils. This was a public form of education that began as an institutional system based on merit despite including nearly the entire population. Indeed, the ratio of students who went onto the upper schools remained at less than 20 percent of the primary school graduates during the prewar period.[11] Toward the end of World War II the gap between advancement based on merit and individual mastery through science education actually widened. Science education, during this time, tended to be viewed as "harmful" to the national project and social integration.

Science education had not only been associated with the social/economic success of individuals of merit, it also facilitated a particular way of establishing relationships between the self and others. Some scientists of education, for example, made critical comments such as: "natural science is a product of human beings. But, once it becomes public, science becomes self-sufficient eventually controlling the people."[12] Such critics were skeptical of a "science" that treated everything (including emotion and spirituality) as inherently operational and controllable.

The "objective" approach to science requires the observer to frame one object in relation to another, to analyze and abstract them in order to extract the substance or find the general rule that results from the process. We can see examples of this in textbooks of secondary education of the period (e.g., "compare differences between the plants placed under [natural conditions and those placed] under other conditions"). Another textbook asks the question, "which golden fish would be most enfeebled [by lack of oxygen] if it were put in three different sizes of aquariums." The observer was then requested to shift his/her objective view into the "subjective" or sentimental point of view that enabled him or her to relate to "the pain of golden fish suffocating."[13] This observation process was incorporated over concerns that if the "subjective" viewpoint were dismissed in scientific learning, it would ignore the "educational" and ethical qualities of human nature (e.g., the sense of togetherness and solidarity). The subjective point of view, as the science reformists suggested, was the primary means for a person to relate to other people. These reformists considered the inclusion of a subjective view of science as a way to enable students to see themselves as part of larger whole. This reformist approach therefore recommended that science education introduce the experimental and subjective approaches simultaneously in order to promote national solidarity. Though these ideas were actively discussed, they were not widely introduced into science education until the end of World War II.

The manner in which science education was discussed (either subjective or objective) thus represented the way in which one should view the self in relation to the universe. The subjective approach to science education, it was argued, provided the best way to ensure seeing one another as organically linked. This approach is called "ecological" in the more recent terminology of the Japanese school curricula. Many of the arguments, at that time, admitted that science education needed to be conducted by the subjective approach because education required that people not only to be individually skilled, but socially ethical. From the opposing perspective, science education was to understand natural phenomena as an external reality for operational and manufacturing purposes. This latter approach enabled one to perceive other objects as controllable and producible. The approach was reasonable and moreover, was accepted by school scientists, who considered science education to be the foundation for technological development as the nation was preparing to enter the War.

Even after the War, these two competing views of science education remained. American influences were obvious in postwar Japan, yet these influences tended only to support one of the two scientific views of prewar Japan. Adopting the problem-solving method, postwar scholars of Dewey seemed to emphasize science education as a way to see one ecologically tied to the rest of the world (i.e., the subjective approach) rather than as a method employed by each individual to discover his/her own answers (i.e., the independent style of learning).

Dr. Vivian Edmiston, who helped the Ministry of Education create the Course of Study of Science, suggested that the preferred method for achieving scientific "rationality" was to introduce it in terms of the interests, concerns, and "desires" of the child, derived from his/her own daily experiences. "The good science teacher will be alert to student progress in using science in the problems that confront him from day to day."[14] Along with these recommendations, science education came to be redefined as being primarily concerned with development of the child's life and experience based on the academic disciplines of natural science and technology.[15] Both approaches to science education thus agreed on the point that learning was best accomplished by the "problem-solving" method as a tool for building a democratic society.[16]

One way to promote a democratic society was to acquire, through science education, the use of the problem-solving approach in the child-centered classroom. This approach de-emphasized contemplation of one's own nature as the starting point of learning, emphasizing instead a more integrated approach to knowledge and understanding. This approach, as applied to the child-centered classroom, relied heavily on Dewey's statement: "I believe that . . . the true center of correlation on the school subjects is not science, nor literature, nor history, nor geography, but the child's own social activities."[17] From this perspective, to teach science was not to teach a finite set of academic topics, but rather to recognize and be a reflection of the child's everyday social world. Thus, the school curriculum was supposed to reflect all areas of the social world of the child, and science is one way of engaging that social world. The problem-solving approach to learning was considered the most appropriate way to achieve this type of integrated curriculum, and best suited to the individual student. Umene was one educator who argued the academic boundary between natural and social sciences was unnecessary and therefore combined natural and social sciences into "social studies."

In postwar Japan, the scientific way of knowing was thus redefined to promote the integration of the curricula and to restructure classroom management, whereas the independent style of problem solving in science education as initially conceived was given little attention. The Japanese concept of social studies arose partially from the critique against science education as an isolated academic discipline.[18] Science education was to be reformed by being incorporated into social studies as an integrated area of academic study. For these critics, the traditional view of science education as an isolated disciplined seemed to represent a premodern version of humanity that differentiated "arts" from "technology." Schooling based on this form of scientism had the tendency to differentiate theory from application, and thus encouraged teachers to cover the basics (theory) before teaching the practical applications to students. It was suggested that if scientism were maintained, it would have a negative impact on all areas of the school curriculum including the humanities and art. Academic knowledge based on scientism served as a system of knowledge that served to perpetuate the division of theory and practice. To avoid this, the reformers argued that one should begin to study what happens in life and the function of the community in facilitating the collaboration of various labors. Referring to Dewey, Umene argued that the "school must not be a studious, but a living place."[19] For Umene, science could be integrated into "social studies" in order to derive knowledge from the practice of life.

Nonetheless, the postwar approach often treated science as a single area of study. Here, science education became the place where the student learned the methods and the legacy of natural science. Schools thus had a responsibility to disseminate the knowledge of science to all. In the study of natural science, the problem-solving method was not a means to satisfy one's own interests and concerns, but a means to understanding the order of physical nature and to utilize it for social wealth and well-being.

This natural science approach emerged from a strong criticism of the Ministry's position that "students raise the quality of life through the acquisition of a scientific way of thinking and acting."[20] The Ministry's goal was to democratize the curricula in order to improve the chances of individual success (in contrast to the traditional merit-base practice of selecting the elite from the student population). This criticism against the Ministry's position was consistent with the criticism of Japanese scholars of Soviet education who saw the curricula promoted by the Ministry as ideologically aligned with the American liberalism represented by Dewey.[21] They could not accept a curriculum based on "individual experiences" because the curricula they preferred regarded the child as a mere unit of the larger whole, a "*homo-social*." Those scholars who defended science education as a single academic discipline desired to emphasize its utility for the entire society, not for the profit of individuals.

The "Sputnik shock" of 1957 strengthened this natural science approach. This event provided some scientists the opportunities to argue that "science education needed to be modernized" and the advanced study of science should be promoted among elite students to improve national competitiveness in the Cold War environment.[22]

Education was severely constrained in both pre- and postwar Japan in the sense that the people were included into different types of the collective formation (i.e., totalitarianism and democracy), but were often excluded within the social system (social strata and positions). Saying that both periods of Japan were ideologically different is legitimate, but it is just as legitimate to say that postwar Japan democracy was still restricted in the form of totality or homogeneity as a nation. This can be called the modern condition in which each individual is connected with others in being immersed in certain types of knowledge (e.g., scientific) in the state. As a result, the manner in which science was taught continued to vary in terms of subjective and objective approaches to inquiry or in terms of specialized versus integrated areas of study. These different approaches to inquiry were responsive to the levels of social mobility that allowed the individual to move within the social system.

Discussing the Individual Governance over Dewey

Dewey was characterized as a great philosopher, psychologist, and educationist, rather than as a "political" thinker in Japan until only recently. But, drawing from the thinking of German social theorist, Jürgen Habermas, Dewey today is being reexamined and discussed in rebuilding the "public sphere" (i.e., community) and public mind of modern Japan.[23] Dewey's pragmatism is viewed in contrast to Cartesian dualism (which has contributed to the production of different types of knowledge that served to maintain the social divisions of labor found in Japanese society). His philosophy of action and experience was, therefore, designed to overcome these Cartesian dualisms. Dewey's approach to develop the individual for the general social well-being is placed in a different cultural system of Japan for collective self-regulation. The individual's problem-solving approach to the world was cast as a governing practice of the self in producing social order. This way of viewing the problem-solving approach to pedagogy advocated by Dewey has recently been used by some Japanese scholars to recast Dewey as a "political" advocate of liberalism.

Recall that Dewey visited prewar Japan while the Japanese version of the progressive movement was taking place. School textbooks also represented a significant shift in the education of children at the time. Textbooks fundamentally changed to emphasizing an "education-adjusted to a child's life" or a child-centered curriculum. This child-centered curriculum's appearance eventually brought the appearance of concern over the depictions of "daily life" in the textbooks. Daily-life portrayals in education were typically associated with a pedagogy that stressed a curriculum that included scenes of daily activity, and further promoted the child as the chief agent of learning. This system of reason shifted the focus of education onto the daily life of the child effectively serving to establish the subjectivity of learning, ensuring that *all* can learn in their own manner.

Oikawa Heiji was a well-known educator of the time who argued for a governing structure for the classroom that recognized the subjectivity of the child, drawing from Dewey's ideas of child centeredness and daily experiences. Oikawa developed his argument by first asserting that the curriculum should consist of a set of continuous activities for fulfilling the nature of a student's learning. Second, it requires well-prepared pedagogical settings, and third, it should be organized and developed around themes focusing upon the common experiences of life (e.g., lifestyle and culture). Each of these features was seen as complementing the other. It was this philosophy of experience that was to form the basis for learning by encouraging a self-learning attitude among children. When the classroom was based on one's daily experiences, teaching and learning could be integrated into the classroom, encouraging the child's own self-directed activity using the child-centered approach. The uniqueness of this experience-based curriculum, according to Oikawa, consisted of both "pragmatic" and "artistic elements."[24] For him, the pragmatic element included opportunities to engage in problem-solving activities that were connected with familiar aspects of living, providing practical opportunities to practice thinking and reasoning.

> The artistic element of the curriculum provided a valuable complement to the pragmatic: There is a different element in life, sentiment, which is not covered by the pragmatic Element . . . [The sentiment appears in the attitudes of] serene contemplation, meditation, peace of mind which is separated from fear, and the joy of a merciful heart toward others, and these attitudes cannot perfectly cognizant by the dualistic view of "theory and practice." Rather, the conceptual position of

"education-as-a-whole" which represents the total view of the individual actions and the whole personality reveals these attitudes.[25]

According to Oikawa, the experienced-based curriculum supported by Dewey's writings includes the "artistic" element of life. The artistic element of life is intended to make the curriculum more extensive by covering the emotional and sentimental aspects of life. Oikawa developed the notion of the "artistic sublime" that integrates emotion and pragmatics. When subject matter consisted of "the legacy of society," the curriculum represented both the sentimental and pragmatic life of the child. He was unique among critics of the dominant approach to education in his denial of the efficacy of teaching directly from traditional textbooks using the customary sequence, preferring instead to restructure the curriculum around the phases of "life" that children pass through.

To be fair, Oikawa credits Eduard Spranger's[26] term, *Gesamtunterricht*, or "whole learning" as an additional source to his idea of the child-centered curriculum largely because Oikawa could not ignore the commitment to German idealism found among many Japanese scholars. Studying the work of Spranger, Oikawa came to see the curriculum as representing the integration of all human conduct, or the totality of human action and experience. This included (indeed presumed) the integration of sentimental elements in life with that of the more academic subjects. The subject matter of the textbooks in use at the time were adapted and integrated into activities based on common (i.e., national) life experiences. The textbooks were useful for this type of learning, particularly since much of the subject matter was already embodied in common depictions of daily life. Oikawa was the primary advocate for incorporating the self-governing aspect of pragmatism to reflect life in terms of collective self-regulation. His major argument was that the self-governance of liberalism was compatible with the need for social control through the use of a problem-solving form of pedagogy.

For some contemporary critics of prewar Japanese education, Oikawa's discussion of a child-centered education was seen as a product of Western influence. They argued that the integration of the curriculum was simply the borrowing and combining elements from both German and American philosophies of education and thus, were not intrinsically Japanese. Others criticized Oikawa for not giving sufficient detail and direction on how to organize the curriculum.[27] Oikawa was challenged to describe more fully the notion of "educated-for-life" and the educative experiences that needed to be provided to students. This challenge was based on the concern for a kind of laissez-faire dependence on the whims of the student. Oikawa was criticized because his student-centered approach to the curriculum did not define the essential elements of "Japanese life." Oikawa's explanation of the child-centered curriculum lacked a focus on the history and geography of one's own native land and culture. This view reflects schooling as a socialization process whether it was the American use of "democracy" or the Soviet use of "industry and fraternity" as organizing features of the curriculum. Similarly, it was argued that the Japanese needed something to uniquely identify their own culture and curriculum.[28] Most of criticisms were, as a result, leveled against the lack of a "national" character that made the curricula uniquely Japanese. Making the curriculum unique means making it "national," an integral part the image of modern Japan. In this respect, Dewey's approach to education was seen as antinational. But, it is fair to say that Oikawa also represented a blending of the Japanese American (Dewey) and German (Spranger) influences.

In prewar Japan, the recognition of the Japanese culture and legacy were considered a major goal for Japanese education. For Oikawa, this recognition of society was realized through the learning process. Even though the pragmatic approach supported a total approach to learning, he sensed the limitation of the pragmatic approach in defining concrete activities and clear procedures. Oikawa argued it was not preferable to think that:

> "Life" could be teachable if it is concerned with the living of the child and adult. [Math] quizzes how much it costs if you buy three pencils for two yen each could be related to living. But, is this type of questioning truly educational? . . . It is not an "education based on life experiences" because the question is not directed by a specific culture of life.[29]

For the recognition of Japanese culture, it was not enough for Oikawa to teach academic subjects along with the "practical" side of living. He eventually called for a "spirit of national independence" that made the goal of education clearer in selecting the learning materials for students.

Oikawa appreciated the idea that students' own interests and needs served as a basis for governing practice. The child's nature was the origin for all activities with which students were to be concerned. But, the argument was always limited insofar as the child-centered curriculum argued that only the interests and concerns of students should form the basis of educative experiences without regard to specific academic content or cultural norms. Although these ideas reflected the essential principles of liberal democracy for Dewey, Oikawa saw them directed toward the preservation of a collective identity. Thus, the child's own nature and interests were to serve as an important means for children to acquire the collective identity of being Japanese. The legacy and culture of Japan, which is equivalent to "Japanese life," were to be transmitted through textbooks in order to maintain Japanese identity and society. This is somewhat different from Dewey's approach to culture and legacy, which was that knowledge can be learned by the individual child through careful scrutiny of the daily activities of ordinary people.[30] Dewey's approach was not, from the beginning, intended to create community as a whole, but to illustrate constructive models of citizenship. Oikawa's approach was to rid the curriculum of a sense of nationalism while realizing some form of collectivism that Japanese people shared. Thus, Dewey's and Oikawa's approaches differed in the manner by which the governing of social identity was created and maintained. These identities were understood in the form of self-regulation for Dewey, and that of individual independence for Oikawa. While the former sought to order the self organically through a transformation in association with the environment, the latter required one to look for the essential individuality of each student.

The postwar democratization of education reaffirmed the will to learn among all children, but was strengthened by the employment of an educational psychology that recognized potential, and more precisely, the desires and goals that each individual may posses. A postwar historian of education noted that "the postwar edition of the Reader was [designed] . . . to take learning and developmental psychology into consideration [as this kind of psychology was] particularly helpful in reorganizing the contents and locations in materials."[31] The use of psychology in postwar Japan was a great help in rearranging textbooks to promote a child-centered curriculum that better reflected the developmental processes of the child. The introduction of psychology into the classroom was therefore intended to help all children participate in learning in the new postwar Japan. This signified a shift in the locus of control from the teacher to the student and thereby transformed the governing structure of the classroom environment.

Postwar Japanese saw Dewey as a sort of *traveling library* in their postwar efforts to establish democracy. An American researcher, impressed by the way in which the Japanese enthusiastically embraced

Dewey as an icon of democracy, wrote: "(T)he Japanese themselves spoke of a 'Dewey's boom' in their country . . . Books on Dewey had brisk sales. A 'John Dewey Night' was celebrated . . . at the University of Hokkaido."[32] But, the way in which Dewey became understood was ambiguous by the way postwar Japanese reconciled themselves with the defeat of the War. For many educators in postwar Japan, Dewey represented educational ideals from the prewar era that could not be fully realized due to the ideological interference of totalitarianism and ultranationalism brought by the imperial regime. Thus, the American ideology was overpowered by the imperial ideology of prewar Japan, and a dictatorial government interrupted the autonomy of the people. However, as with others that have studied and attempted to understand the philosophy of Dewey, the postwar Japanese elaborated multiple interpretations of his work by trying to inject his ideas of the "modern" (i.e., new and advanced) into the premodern system of Japanese education. For some, a Deweyan approach stressed education for citizenship, based on individualism and liberalism. For others, studying Dewey became an opportunity to make Japanese more aware of the differences between Japanese and Western traditions and values.

Dewey was also influential in the process by which Japanese intellectuals implemented the recommendations of the *Report of the American Education Mission to Japan* of 1946. A group of twenty-six intellectuals presented what they thought the Japanese people most needed to learn from the Report in 1950. The group concluded that it was urgent and practical to focus attention on how the education system needed to be developed and modernized, rather than on how education's political and administrative system was to be transformed.[33] Thus, the twenty-six Japanese intellectuals agreed to address the needs of education rather than the supra-structure (i.e., political and ideological structure) of education. The majority of the members firmly believed that the "modernization" of education involved allowing individuals to explore their potential and to become "independent" in order to change the country in the desired direction. Hence, teaching needed to move from a focus on what students should learn to what they wanted to learn in order to make the country democratic. Founded upon Dewey's theory of individual differences of needs and interests, this child-centered approach required an individual to know oneself as independent (i.e., different) from others. Nonetheless, the child-centered approach was largely perceived as a pedagogical approach primarily intended to simply encourage an appreciation of the unique nature of each child. As a result, many considered the child-centered approach as merely a strategy for promoting individual growth rather than a political strategy for promoting social reform.[34]

Among educators, the child-centered approach was understood as a method for "learning from experiences." Social studies in postwar Japan represented a particular field of study to convey a set of knowledge drawn from an experimental approach to classroom activity. The "experimental-based curriculum" was designed to utilize the "project method," a familiar pedagogical approach that had been used in prewar Japan. The project method and experience-based approach were largely attributed to Dewey, though its implementation in Japanese schools often looked quite different than the original examples offered by Dewey. Nor were the Japanese themselves in complete agreement on what an experience-based approach should look like.

The Japanese social studies program was intended to stress the development of a student's disposition and attitude that would contribute to the construction of a new society. According to the program, students were expected to "expand social experiences" through social studies rather than gain abstract knowledge based primarily on memorization, as was common in prewar subjects such as History and Ethics.[35] The new program encouraged teachers to develop an awareness of concrete themes surrounding the lives of students and stimulate their creativity in addressing those themes.[36] These themes were to focus on real-world problems. The program defined this "principle of 'learning-by-doing' as the most important aspect of studying social studies." In general, the phrase of "learning-by-doing" was taken as an example of Dewey's approach to education.

The task of developing the details of the social studies curriculum fell to a group of Japanese educators (selected as the members of the *Core-Curriculum Association*). In attempting to work out the details of the social studies curriculum, a debate soon emerged over just what the "core" of the social studies curriculum should consist of. For some, the core represented a "learning by living" approach, where the subject matter content was based on the stage of development and normal activities of life that students should be engaged in. This was seen as an education focused on developing primarily an individual sense of identity. For others, it was believed that social studies should first and foremost be about the development of a collective identity. For these educators, the curriculum needed to focus primarily on the rights and responsibilities arising from "citizenship." It should be noted however that both groups struggled over the concept of identity formation—whether individual or collective—in a similar manner to that debated in prewar Japan. A good example of this modern influence is the persistent continuity of history that continually evokes the same framework for targeting the "problems" discussed by members of the society. Very few examined Dewey's pedagogy as a self-governing system designed to integrate the collective and individual regulation of society. Instead, most of Dewey's supporters among the Japanese (whether pre- or postwar) treated pedagogy as a means for promoting independence.

In the ensuing debates, Dewey was interpreted by various groups in a manner supportive of their positions. Umene, for instance, who regarded "daily living" as the core of social studies, legitimized this way of thinking by defining the curriculum using Dewey's phrase, "learning through, and in relation to living."[37] By such references to Dewey, Umene and other members of the Association attempted to bring back much of the prewar "experience-based" curriculum. For proponents of a more citizenship-based approach to social studies (e.g., Mori Akira), the importance of a collective identity and common knowledge were highlighted as key elements of Dewey's thought. They also explained the key differences in the positions taken toward social studies by referring to Dewey: "The difficult course we have to pursue is to recognize education as generating 'genuine learning' and purposeful activities simultaneously."[38] "Genuine learning" is completed by the individual adopting the experienced-based approach, while purposeful activities were realized through a focus on learning for citizenship. Mori interpreted "learning from experiences" as a situation where the learner understands the external conditions in an individualistic sense, which he called "subjective reality." But, he argued that there also exists another reality, which he characterized as "objective reality." This sort of reality could be different from the subjective reality understood by the learner. Education, Mori argued, was supposed to bridge the chasm between these two different realities. Education had to create a way for the learner to assimilate a common set of knowledge into one's own personal understanding. Therefore, problem solving was seen as a powerful pedagogical approach for this connecting of the learner's individualistic understanding with "objective" reality. Mori's approach was to define "experience" as consisting of both a personal and collective component. However, once he pointed out this collective aspect to experience, historical and traditional knowledge could no longer be ignored.

Mori's view of education suggested that education was viewed as a major production of modernity and served to function as a bridge between two different worlds, private and public, or self and other, in post-feudal societies. The boundary between private and public was not explicitly clear in the feudal system of Japan. Education was, thus, burdened with the role of sublimating the two worlds. The problem-solving approach as a scientific rationale seemed beneficial for the sublimation of the two, but the discussants varied in their opinions of just what problems were to be worked on.

Some Concluding Thoughts

Although Dewey's ideas were critically examined, he was often positioned on one or the other side of the dualistic views advocated by the Japanese, depending upon the perspective of the person interpreting Dewey. He was regarded as nonscientific from the hard-science viewpoint, but extremely practical from the traditional viewpoint. The varied ways in which he was characterized were largely the result of the social and historical perspectives of the Japanese. Dewey as an *indigenous foreigner*, however, contributed to the constant discussions surrounding Japanese nation building and the mobilization of the population toward the modern/global project. Dewey was thus useful as a *traveling library* for discussing both inclusion (i.e., to consolidate as the people) and exclusion (i.e., to socially divide into labors) among the Japanese. As another example, the problem-solving approach (the pragmatic way of learning) was influential in discussions designed to unify the way people live or individualize their unique way of being. Japanese scholars and educators not only read his works, but also obtained the chance to revise the national and social formation of their own country through the use of Dewey's ideas.

The problem-solving, child-centered, and experience-based approaches are all related to Dewey's pragmatism. These approaches enabled the learning of democracy by all. This democratic ideal is reflected by the fact that every member of society has the opportunity to join the learning community (school), and learn by problem-solving strategies applied to individual experience and thereby find solutions for one's own understanding. However, this approach was not widely adopted in Japanese schools, though it had been defined as ideal during both periods of Japanese education. While the democratic approach to learning utilizes positivistic procedures through the natural inquiries of individuals, it also contains the risks that learning may become individually isolated and independent from the understanding of others. This binary sense of benefit and risk through learning is related to the points made in Popkewitz's introduction regarding the idealism of bringing together Calvinist themes of redemption and the political forms of a participatory democracy. Dewey was convinced of the value of participatory democracy through the exercising of individual acts, though the Japanese often saw this as merely an ideal.

Because the individual act was endorsed by problem solving and a scientific attitude, "science" was not viewed as the technology for production but rather as the technology by which to govern oneself in society. Those Japanese who discussed Dewey's pragmatism seemed not to realize the scientific sublime in Dewey's thoughts that embodied individual acts in concert with a participatory society. For the Japanese, science was an "advanced" form of knowledge that was "disenchanted" from the traditional approach to knowledge and thus unburdened the Japanese from the past (especially with respect to the defeat of the War). The memory of the past caused the Japanese to perceive science in a manner different from pragmatists like Dewey. They considered science to be mostly a means to enter the world stage as a modern society.

As demonstrated in this chapter, the Japanese pursued the "modern" through their own ambiguous and contradictory understanding of Dewey. His thought contributed to mobilizing the population toward democracy, and helped to rid them of the totalitarian bondage of Imperial Japan. Through the eyes of the Japanese, the modern was extremely ambivalent in its implications for nation building and self-realization. This ambivalence of the modern as problematic was consistently repeated in both prewar and postwar Japan, while the country itself transformed its national ideology from imperialism to democracy. Ideologically speaking, Japan had changed in many profound ways, but its mentality maintained the same dualistic and ambivalent sense of identity.

Notes

*   Japanese names throughout this chapter, except those of Japanese Americans, appear in order of family name first.

1. Saburo Ienaga, "The historical significance of the Tokyo trial," in *The Tokyo War Crimes Trial: An International Symposium*, ed. C. Hosoya et al. (Tokyo: Kodansha Ltd. and Kodansha International Ltd., 1986); Masao Maruyama, *Koei no ichikara* (Tokyo: Miraisha, 1982).

2. William G. Beasley, *The Rise of Modern Japan* (New York: St. Martin's Press, 1990).

3. Robert S. Schwantes, *Japanese and Americans: A Century of Cultural Relations* (Connecticut: Greenwood Press, 1976).

4. Monbusho, *Shogakuseito-kokoroe* (Tokyo: Normal School, 1873). *http://www.tulips.tsukuba.ac.jp/exhibition/bakumatu/sihan/seitokokoroe. html* (accessed July 22, 2004).

5. See the discussion in Victor N. Kobayashi, *John Dewey in Japanese Educational Thought* (Ann Arbor: Malloy Lithoprinting, Inc., 1964), 37–50.

6. Akihiro Mori, "Dyui kyoikugaku kenkyu no kadai" *Kiyo* (October 1961): 59–65.

7. John Dewey, *School and Society / Child and Curriculum* (Chicago: The University of Chicago Press, 1990); Satoru Umene, "Seikatsu tangen to Mondai tangen," *Karikyuram* (April 1950): 1–6.

8. Monbusho, "Jinjo shogaku rikasho," in *Rika kyouiku-shi shiryo, vol. 2,* ed. Kiyonobu Itakura (Tokyo: Toho, 1911).

9. Monbusho "Shogaku rikasho ni taisuru iken hokoku," in *Nihon no rika kyoikushi,* ed. S. Hori (Tokyo: Fukumura, 1913).

10. Gentaro Tanahasi, "Kaitei shinrika kyoujyu-ho," in *Rika kyouiku-shi shiryo, vol. 1,* ed. Kiyonobu Itakura (Tokyo: Toho, 1918).

11. Monbu-Kagakusho, *Wagakuni no kyouiku tokei* (Tokyo: Zaimusho Insatsu kyoku, 2001).

12. Isaburo Kanbe, "Wagakuni genji no rika kyouiku toumen no jissai mondai," in *Rika kyouiku-shi shiryo, vol. 1,* ed. Kiyonobu Itakura (Tokyo: Toho, 1936).

13. Eiji Sato, *Senjiki no kyoukasho ni okeru rekishisei no mondai,* paper presented at the meeting of Japan Society for Curriculum Studies, Aichi (2004).

14. Vivian Edmiston, "Newer methods of science teaching," *Kagaku to kyouiku* (October 8, 1947): 1–3.

15. Kiyonobu Itakura, *Nihon rika kyouikushi* (Tokyo: Dai'ichi houki shuppan 1968); Satoru Umene, "Seikatsu gakko to koa-karikyuram," *Karikyuramu* (January 1949): 5–8.

16. Kiyonobu Itakura, *Nihon rika kyouikushi* (1968); Nisseiren, *Kodomo no seikatsu wo hiraku kyouiku* (Tokyo: Gakubunsha, 1988).

17. John Dewey, "My pedagogical creed," in *John Dewey on Education: Selected Writings,* ed. R. D. Archambault (Chicago: University of Chicago Press, 1964); Hiroshi Usami, "Dyui no kyouiku houhou-ron to sono eikyo," in *Nihon no sengo kyouiku to Dyui,* ed. S. Hiroshi (Tokyo: Sekai Shiso sha, 1988).

18. Satoru Umene, "Shin kyouiku to konpon mondai," in *Umene Satoru kyouiku chosakushu* (Tokyo: Meiji Tosho, 1948).

19. Satoru Umene, "Shin kyouiku to konpon mondai."

20. Monbusho, "Gakushu shido youryo / Ippan hen: Shian," in *Rika kyouiku-shi shiryo, vol.1,* ed. Kiyonobu Itakura (Tokyo: Toho, 1951).

21. Koji Tanaka, "Sengo kyouiku no mondai to Dyui no eikyo," in *Nihon no sengo kyouiku to Dyui,* ed. H. Sugiura (Kyoto: Sekai shiso sha, 1998).

22. Kiyonobu Itakura, "Minshu shugi no hata wo takaku kakageyou," in *Rika kyouiku-shi shiryo, vol. 1,* ed. Kiyonobu Itakura (Tokyo: Toho, 1963).

23. Manabu Sato, "Kokyoken no seijigaku," *Shiso*, (January 2000): 18–40.
24. Heiji Oikawa, "Shogakko karikyuramu no kaikakuan," *Kyoiku* 3, no. 9 (1935): 57–67.
25. Heiji Oikawa, "Shogakko karikyuramu no kaikakuan."
26. Spranger was a student of Wilhelm Dilthey and visited Japan with a cultural mission from Germany in 1936.
27. Yonekichi Akai, "Karikyuramu no kaikakuan wo yomite," *Kyoiku* 3, no. 9 (1935): 50–53.
28. Chikuji Kinoshita, "Shogakko karikyuramu no kaikakuan ni tsuite," *Kyoiku* 3, no. 9 (1935): 46–49.
29. Heiji Oikawa, *Kyoiku* (1935): 30.
30. John Dewey, *School and Society / Child and Curriculum*.
31. Tomitaro Karasawa, *Kyokasho no rekishi* (Tokyo: Sobunsha, 1956).
32. Victor Kobayashi, *John Dewey in Japanese Educational Thought* (1964): 1–3.
33. Sei'ichi Miyahara, "Kako no nihon no kyouiku no kekkan" in *Amerika kyouiku shisetsudan hokokusho yokai*, ed. H. Sugo et al. (Tokyo: Kokumin Tosho Kankokai, 1950).
34. Hajime Tamaki, "kyouiku to nihon no seiji taisei," in *Amerika kyouiku shisetsudan hokokusho yokai*.
35. Monbusho, "Gakushu shido yoryo shakaika-hen: Shian," in *Shakaika kyouikushi shiryo, vol. 1*, ed. K. Ueda (Tokyo: Horei Shuppansha, 1947).
36. Monbusho, "Gakushu shido yoryo shakaika-hen: Shian," 199.
37. Satoru Umene, *Karikyuram* (April 1950): 1–6.
38. Akira Mori, "Keiken to kyouiku," *Karikyuramu* (August 1950): 11–14.

Index